Identification and Management of Horticultural Pests

NIPA® GENX ELECTRONIC RESOURCES & SOLUTIONS P. LTD.
New Delhi-110 034

Identification and Management of Horticultural Pests

A.M. Ranjith
Professor, Department of Entomology
College of Horticulture
Kerala Agriculture University
Thrissur, Kerala, India

NIPA® GENX ELECTRONIC RESOURCES & SOLUTIONS P. LTD.
New Delhi-110 034

NIPA® GENX ELECTRONIC
RESOURCES & SOLUTIONS P. LTD.

101,103, Vikas Surya Plaza, CU Block
L.S.C. Market, Pitam Pura, New Delhi-110 034
Ph : +91 11 27341616, 27341717, 27341718
E-mail: newindiapublishingagency@gmail.com
web: www.nipabooks.com

For customer assistance, please contact
Phone: + 91-11-27 34 17 17
Fax: + 91-11- 27 34 16 16
E-Mail: feedbacks@nipabooks.com

ISBN: 978-81-19215-09-6

Composed and Designed by NIPA®.

PREFACE

It has bean a long journey into completing this book. The book is intended for the undergraduate and post graduate students of Agriculture, and Agricultural Entomology. As has been discussed in the introduction, this book is probably the only comprehensive book with photographs and most recent scientific names, along with the old names, which has been published in the last forty years. I take pride in it, and attributes it to the continuous pestering (for good reason) of me for recommending to them a text book that would be useful for them in the field, in the class, and for JRF and other comprehensive exams. Even though a book of such a magnitude cannot be foolproof, (there are still some typographical errors, front problems and such minor errors) that the guide for the scientist, the teaching faculty and even policy makers, who would take time off to get informed on the ground realities of pest and pest management. I place this book for all these groups.

There is something more to add. In Kerala, the use of Endosulfan has been banned for a long time. The other insecticides that have been banned in Kerala are Carbofuran, Phorate, Methyl, Parathion, Monocrotophos, Methyl, Demeton, Triazophos and Profenophos. None of the recommendations in the book, for use of these pesticides in crops, should thus be used in Kerala, it is also to be noted that carbaryl is no longer available, and will hence have to be substituted with other suitable insecticides. These hard chemicals should be substituted with suitable newer chemicals, which are more environmental friendly, ecologically sound and economically cost effective. Indeed, the recommendations are just for guidance, and as newer strategies for IPM develop, the use of chemical insecticides is bound to decrease.

The users are requested to bring to my notice, any corrections that they come across in the book by e-mail to yourfoundation@gmail.com; or ranjith.am@gmail.com

A.M. Ranjith

CONTENTS

SECTION–III: ORNAMENTAL PLANTS

SECTION–IV: PLANTATION CROPS

SECTION–V: SPICES

SECTION – VI: TUBEROSE

SECTION – VII: VEGETABLES

INTRODUCTION

This book has been written, primarily due to my own felt need to have a comprehensive text book on the pests of horticultural crops and methods to manage it, in the integrated way. For a long time, I had felt this, and also had to search a lot of sources to get an idea about the adults, their photos and methods to reduce their attack. Chemical control in its entirety is not intended, but a general indication has been given, because many states have many leading chemicals, being sold locally and hence the brands and recommendations differ. But the pests, by themselves, and their biology is always the same. It was thus more of a need to have a text book to teach the UG and PG students, that this book was written. It was also necessary that any practicing field level worker like officers of the state department of Agriculture and commodity boards, and every practicing farmer with an ability to read English text book, should be guided into the basic facts about the pests of crops. This is also necessary to avoid confusion between pests and predators, especially the coccinellids or lady bird beetles which can be either useful natural enemies or pests of crops. Hence, a photograph would go a long way in avoiding confusion and help to decide whether to keep the insect or manage it by IPM practices. By limiting the objective to this, a detailed reference section is given in a book purely for the students, becomes unnecessary and hence, this book does not have an authentic reference section .The reader has many sources to get the latest reference, in detail, by collecting it from the internet and other sources. It would have also made this book much more bulky and costly, but without really serving the purpose. Hence, the reference section has been deliberately avoided.

There is also an arrangement, based on fruits, medicinal and aromatic plants, ornamental plants, plantation crops, spices, tubers and vegetables. In describing the pest and in the usage of the scientific names, the emphasis has been on the usage of the latest nomenclature, but wherever feasible, the old name has also been incorporated. This is because for the persons who do not use the latest names and do not, as matter of routine, get updated, the old familiar name is the only way of knowing about the pest. It thus becomes necessary to keep the older, but most familiar names also. Many a time, it is our experience that the older name comes back after a few years, when a fresh review is made on the scientific name. Hence, it is more prudent to use the time tested more familiar scientific names, but also use the latest names. An example is the usage of newer pheromones, and other behavior modifying chemicals, where fresh, effective chemicals are invented and popularized, making earlier technology and products obsolete. Even the methods of dispensing the semiochemicals or other behavior modifiers get changed in time, paving the way for more efficient, effective methods of management. Just as it was, that we used the term insect control earlier, now the term that we use is insect pest management. Controlling the insect is never a necessity for the practicing entomologist, and it is actually, contrary to our goals. Our aim gets limited just to manage the pest so that we get an efficient, cost effective, non-polluting, fresh, uncontaminated plant produce, which has to be healthy and nutritiously rich for the end user. Goals and priorities have also changed, so that the current idea is to leave a pest residue, and to support a natural enemy –pest- tolerable crop damage system that would ensure the suppression of the pest 24 × 7 × 52 system that is most effective. All other systems have their limitations in the 365 day milieu.

It is also necessary for the student and also the practicing entomologist to know more about the important pests of crops as also the minor ones. There is a description of almost all the pests in any crop, including the most minor ones. The student should not get confused with this. In the study of entomology, it is important that we concentrate on the major, endemic and established pests. These would be limited to a handful or even one or two in many crops. Others become problematic only once in a while, usually, once in 10-12 years or so. But the detailed knowledge of the pest in a crop is needed in the case of major pests only. Again they differ between regions and states, primarily due to the prevailing weather conditions in the area. Thus there are dry season pests, wet weather pests and pests that hibernate, pests that are serious in kharif or rabi, pests of winter, pests that are managed just after the emergence of the first season immature after winter, and many other inclement factors. It thus becomes evident that a student has to be aware of the biology of the pest, and the weakest link in its biology. This cannot be done for all the pests, and hence the importance of the major pests. However, the book will never be comprehensive if the sporadic pests are not described. It will be seen that their description is limited to a few words only. This is just to expose the reader to the presence of the pest in the crop. When once in a while the pest makes its appearance, it would only be prudent to get the pest identified with the help of the description, and then to make a thorough search elsewhere, so that, a comprehensive knowledge is gained on it. It is most likely that most of the information is available in this book itself, as pest of some other crop. It will then be possible to read on the pest in that chapter, and manage the pest effectively using the latest techniques available to the crop management expert.

A new set of crops comprising many of the medicinal and aromatic plants have been included in the book. The chapter is probably the first of its kind in a text book of Entomology. What is more, the crop pests have been photographed, and also, the photo of the crop itself is given, so that some of these crops, which are known by other names, can be easily recognised. These crops are by themselves, insect deterrent in many cases, where farmers use them as repellants. But they also harbour specific pests. This is one area which the student has to bear in mind. No crop is free from insect attack, and usually there are specialist pests that attack even the most accepted toxic plants. The typical case is the pests of tobacco and neem. The very fact that they are toxic to humans or other pests makes it an ideal host for many specialist pests. This way, they avoid competition from most of the other pests for utilizing the plant as a host, so that interspecific competition for food and shelter is minimal. The medicinal plants include ambrette, aswagandha, *Coleus aromaticus*, davana, fennel, french basil, isabgol, lemon grass, mentha or mint, opium, patchouli, periwinkle and sarpagandha and their pests have been described.

In the following pages, an attempt is made to familiarise the student on the major pests of crops, which they have to concentrate. Others assume significance only occasionally.

The first crop that is being dealt with in the alphabetical sequence is almond. The important pests include the San Jose Scale, *Diaspidiotus (=Quadraspidiotus) perniciosus,* Green peach aphid/leaf curl aphid, *Brachycaudus helichrysi* ,Tent Caterpillar, *Malacosoma disstria,* Apricot scale /European Fruit Scale, and the *Lecanium (Arthenolecanium) corni*

.Scales and aphids are usually managed in their period of least occurrence, and when they are vulnerable. For this reason, use of oils is good in such crops where the plants have the overwintering mechanism of shedding of leaves and sprouting in spring.

Amaranthus is attacked by the leaf caterpillar or webber, *Spoladea (Hymenia) recurvalis,* but hymenia is still the popular name. The other leaf webbers, *Psara basalis*, and *Eretmocera impactella* are also important. In all these, monitor regularly and mechanically remove the first brood. Heavy attacks can be prevented by use of botanicals. During dry weather, thrips and mites cause significant damage.

The medicinal ladies finger or kasturi venda/bende/bhendi being closely related to the common bhendi, is having the same pests like the shoot and fruit borer,aphids and mites. Management is more by mechanical control and / or botanicals. Pesticides are not recommended.

Anthurium, the ornaments plant, has the thrips and white flies as occasional pests, but the most important problem may be the snails in the greenhouse. Regulated watering and preventing moss growth is the answer to this pest.

Amla, aonla or Indian gooseberry has the gall maker Betuosa as one important pest affecting the growing point. The fruit borer, Deudorix is also important.

In the case of apple, the major pests include the apple codling moth, *Cydia (Carpocapsa) pomonella,* The San Jose Scale, *Quadraspidiotus perniciosus,* the woolly apple aphid, *Eriosoma lanigerum* and the cottony cushion scale, *Icerya purchasi*. All of them are important and also international pests. In India as also in the history of biological control, the classical case of biosuppression of the cottony cushion scale stands aloft as the wonderful, classical, pioneering success of biological control of any crop pest in the whole of the world. This is also the history of setbacks while using the most predominant natural enemy and the success of a small, but sturdy, adapted natural enemy in another land. The student is advised to read further on the topic, which gives an insight into what a native natural enemy does in its natural habitat and how it may fail in a new environment; it also demonstrates how a very inconspicuous, unassuming predator can work wonders in a new environment, this time from Australia to California in America. Other pests have been bad in most places, and the San Jose scale is a typical case of a pest being controlled in the weakest link in its life cycle, and also the use of oils for pest control, still one of the oldest strategies, successful over decades. However, the mineral oil formulations have changed for the better, and we have different forms of oils, tailor made for specific uses. In apple, thrips, whiteflies, mites and snails are also important to a lesser degree. While the mites and thrips cause quality reduction, especially for the flower, the snails are problematic for the plant.

In the case of Apricot also, the San Jose scale is an important pest. The other pests to be careful about are the Indian gypsy moth: *Lymantria obfuscata* and the Apricot chalcid: *Eurytoma samsonowi.*

In Arecanut, the Spindle bug, *Carvalhoia arecae*, the root grubs; *Leucopholis burmeisteri* and *Anomala varians* and the inflorescence caterpillar, *Tirathaba mundella* are important. Most of the crop loss can be attributed to the spindle bug but general debilitation of the plant is also due to the root grubs, especially in the loose sandy soils. Management strategies for the root grub involves adult capture, and treatment of the first instars, when their numbers are more, but confined to the top layers, and easily manageable by dust or EC formulations. The inflorescence caterpillar is an occasional pest, but usually an unnoticed enemy, to be careful about in endemic areas.

In the case of aroids and yams , the mealybugs, *Pseudococcus citriculus, Rhizoecus* sp. the scale insect, *Aspidiotus (Aspidiella) hartii,* and the arecanut beetle, *Araecerus fasciculatus* could be problematic. All of them are pests usually only in storage, even though a neglected crop can get the infestation started in the field. Better farmers do not face an insect problem generally.

Mealy bugs, and cow bugs may cause damage to Aswagandha, but epilachna beetle and sphingids during sporadic attacks may be more troublesome.

Banana is the next in the crop list. The major pests are the Pseudostem borer, *Odoiporus longicollis*, the rhizome weevil, *Cosmopolites sordidus*, and the banana aphid, *Pentalonia nigronervosa.* The aphid is important only as vector of the bunchy top virus .Occasionally, the leaf caterpillar spodoptera and pericallia have also been seen as pests, but is either mechanically collected in the early stages, or a simple non residual spray given on the affected leaves, to prevent the growth and spread of the pest. Here is the importance of constant vigil and monitoring by the farmer. As I have indicated already, pest management strategies are constantly changing. One example here can be the use of tissue culture banana for rhizome weevil management, since, there is no corm and the corm weevil cannot attack the corm while planting. It is not carried over from the corm as in sucker planting. By the time the corm develops, the plant is ready for harvest. The tissue culture plant in small bags also offers convenience in treating the plant for disease and pest management.

The crop Ber is attacked by the Ber fruit borer, *Meridarches scyrodes*, which is an important pest in south India. Other pests are the ber fruit fly, *Carpomyia vesuviana* and another fruit fly *Bactrocera zonata.*

Betelvine has two major pests, the betelvine bug, *Disphinctus politus* and the scale insect, *Lepidosaphes* sp.

Bhendi or ladies finger is the next in the crop list. Here, the major pests are the shoot and fruit borers, *Earias vitella (=E.fabia)*, *E. insulana*, and the Leafhopper or green Jassid, *Amrasca biguttula biguttula.* Occasionally, the fruit borer, *Helicoverpa armigera* and the semiloopers also become pests. However, management is primarily based on the fruit borers, and use of either botanicals or mechanical control is the only option once harvesting has started, since the fruits are harvested almost as matter of routine once in two or three days.

Brinjal is next, and the pests of importance are the shoot and fruit borer: *Leucinodes orbonalis*, the epilachna beetle/hadda beetle; *Henosepilachna viginti octopunctata,* brinjal lacewing bugs; *Urentius hystricellus,* and the red spider mite; *Tetranychus telarius* (*T. urticae*) *fruit borers and epilachna are endemic,* and the other two are pests towards the later stages, especially in summer and when the crop has had some initial pickings.

The cruciferous vegetables, cabbage, cauliflower and others have many pests in common and could be dealt with as a single crop. Cabbage and crucifers are most commonly attacked by the Diamond back moth. The name has come from the diamond shaped areas that are seen on the adult while it is resting. The larvae are defoliators and initial monitoring and management as the key to IPM involving minimal loss. Growing trap crop of mustard all around, spraying the trap crop, use of pheromone traps, parasitoids and predators. Use of *Bacillus thuringiensis*, and its formulations and finally resorting to chemical spraying if the attack goes beyond control are the strategies for managing this very important pest. We also come across many other lepidopterous larvae including helicoverpa, spodoptera, *the* cabbage borer, *Hellula undalis,* the cruciferous leaf webber *Crocidolomia binotalis,* and *Trichoplusia.* The key to IPM is to see all these pests in totality so that when many of them are present in various pockets and in isolation, a total spray might well manage all.

But there has to be specificity in using NPV, for helicoverpa and spodoptera. Identify the pest and use its bioagent, to be successful. Remember, that in biocontrol, most of them are very specific. Know the biology and manage. For cutworms like Agrotis and others, identify them, and manage, since they are not seen during day. Cabbage also has one hymenopteran pest, the mustard sawfly, so named due to its saw like ovipositor. It is usually a minor pest, just like the cabbage aphids, Brevicoryne *brassicae* and mustard aphid, *Lipaphis erysimi*. Occasionally in the plains where tropical tolerant crucifers are being popularised, the leaf eating caterpillar, *Spodoptera litura* has become problematic.

The queen of spices, cardamom is the next crop whose important pests have to be familiarised. Cardamom has many pests, but the most economic loss is caused by Cardamom thrips; S*ciothrips cardamomi;* Shoot, panicle and capsule borer: *Conogethes (Dichocrocis) punctiferalis* and the root grub, *Basilepta fulvicorne.* Even in this, the thrips is a major enemy, which causes corky growth on the husk and loss of aroma, which causes a lot of economic loss. The borer is also problematic, and in some instances, the root grubs, along with the rhizome weevil also may cause damage to the underground portions.

Carnation, one of the pleasant ornamental crops, is being attacked by two pests, the aphids, *Myzus persicae* and also the red spider mite, *Tetranychus urticae*.

Cashew, the next crop, has two major pests, The Tea mosquito bug, *Heliopolis antonii* and the stem and root borers, *Plocaederus ferrugineus* & *P. obesus*. The tea mosquito, so called because it attacks tea flushes, is one of the most important pests of cashew. It reduces cashew yield very significantly, due to increasing populations that attack the new flushes, then the inflorescences, and then the developing nuts. The stem and root borers are not as widespread and rampant as the tea mosquito, but is a silent killer of odd tress and branches.

Cassava or tapioca, the common mans food, has some pests like the Cassava scale; *Aonidomytilus albus*, attacking the setts, the termite: *Odontotermes obesus* attacking the newly planted setts in dry weather and also the dry crop; the white grub: *Leucopholis coneophora,* and recently, the spiralling whitefly,*Aleurodicus dispersus*. The spider mites, *Tetranychus cinnabarinus* is always present during summer, under the leaves, but does not cause significant yield loss.

Thrips *Scirtothrips dorsalis* and *Thrips palmi* are one of the major pests in Chilli and capsicum. Initially they are pests of foliage, causing crinkling, but later attack the flowers and cause flower drop or fruit malformation. These are typically summer pests, and can be managed by repeated spraying or cooling of plants with water, but in persistent attack, spray low residual insecticides. The spiralling whitefly/cottony whitefly; *Aleurodicus dispersus* has also become a nuisance on the undersurface of the leaves, reducing yields and demanding chemical control. This is the pest of the current decade, and also should be managed by irrigation, use of botanicals or soap spray and later by insecticides. The broad mite/ yellow mite, *Polyphagotarsonemus latus* and other mites also are pests, especially in dry weather. Attack by mites is widespread, and so diagnose the pest correctly, and when soap sprays and botanicals fail, go for specific acaricide. Other pests include caterpillars of spodoptera, helicoverpa and cutworms, which can be managed by monitoring, timely sprays and later by specific low residual insecticides. The tobacco caterpillar; *Spodoptera litura* and fruit borer; *Helicoverpa armigera* are both leaf feeders and fruit feeders.

Chinese potato or koorka is not a significant crop in many states, but central Kerala relishes the tuber, but the pests are never at high levels to demand management. This text is probably the only one that has a good description of the pests, and nematodes are the major enemy. Mulching with *Hyptis suaveolens* or *Lantana camara* or neem leaves or other retardant plants before planting can be a good strategy for management.

In China aster, there are few pests like the chrysomelid beetle; *Aulacophora foveicollis,* the flower caterpillars; *Helicoverpa armigera,* the aphids ; *Aphis* spp. and also the root aphid; *Macrosiphum* spp. Chrysanthemum harbours some pests like the aphids of different genera ; *Macrosiphoniella sanborni; Myzus persicae; Brachicaudus helichrysi; Aphis gossypi;Aphis fabae;Aulocorthum circumflexum.* All these usually attack the young growing parts and cause crinkling of buds and flowers. The other important pest is the thrips species belonging to *Thrips tabaci, Thrips nigrophilosus, Frankliniella tritici, Hercinothrips* sp. and *Microphalothrips abdominalis*.

The Aphids, *Macrosiphoniella sanborni; Myzus persicae; Brachicaudus helichrysi; Aphis gossypi; Aphis fabae; Aulocorthum circumflexumi* are all pests of chrysanthemum, but Macrosiphoniella is the most predominant. Leaf miner and flower beetles also pose problems.

Cinnamon butterfly and blue butterfly as larvae are the most important pests of the crop. The crop also has some unique pests, and the reader is directed to go to this chapter and read it.

Pests of citrus are a must read, because of the many pests like the citrus butterfly, fruit sucking moths, leaf miner, black flies , psyllids, scales and the cottony cushion scale, which are unique as pests, historical Iceria biocontrol success, and as preferred topics for question paper setters.

Except the stem borer, affecting the young seedlings and young plants, the clove tree does not get affected by many pests. The attack by the sucking pests like scales, thrips and mealy bugs are not significant. On few occasions, the psyllid causes galls on the leaf, but this is also not significant. Clove is best grown without insecticides.

Cacao Mealy Bug, *Planococcus lilacinus* is the most important pest in cocoa. The bark borer, Zeuzera may pose problems in individual trees. The mealy bug infests all tender parts of the cacao plant such as shoots, terminal buds, cushions, cherelles and pods and cause considerable damage to all the affected parts. In heavy attack, insecticidal control may be needed. Addition of teepol or soap is needed for spreading of the insecticide and to dissolve the waxy coating of the mealy bugs. The tea mosquito bug is a pest of new flushes. For the cocoa farmer, this book gives the most comprehensive account of the pests of cocoa and their management.

The chapter on coconut is again, very elaborate, because of the vast experience of the author in coconut entomology. Nevertheless, the important pests could be limited to the red palm weevil, being easily managed by the pied piper technique of pheromonal management, proposed by the author. This caution of leading the pest outside your farm is a must, or will cause more damage if incorrectly handled. A new problem is the eriophyid mite affecting the buttons and nuts, and should be systematically managed. Rats do pose a problem, and do not come under the purview of this book but has been included for the help of the farmer. Root grubs, coreid bug and mealy bugs are also problematic.

The major pests of coffee are the White stem-borer, *Xylotrechus quadripes*, the coffee berry-borer, *Hypothenemus hampei* and occasionally the red borer, *Zeuzera coffeae*. Severe infestation by the berry borer may result in heavy crop loss.

Medicinal coleus is sometimes affected by the black lacewing bug, but never is managed beyond mechanical collection and destruction.

Coriander, cumin and fennel are affected by the coriander aphid:

Hyadaphis coriandri, and the green peach aphid, *Myzus persicae.*

Cowpea and other pulses are dealt with as one, since the pests of these are mostly common. They have many problematic pests and the student has to learn them, their biology and management, since these pests limit production, both as vegetables and even in field pulses. By far, the most important are the fruit borer complex that limits production to the maximum. Aphids, and in recent past, the papaya mealy bug has also assumed importance, but can be managed by insecticides, in severe cases. Many types of pod bugs and the epilachna beetles also reduce the yield.

Crossandra is a minor crop and has some scales and other minor pests affecting it.

Cucurbits are important vegetables, and are always attacked by the fruit fly. The best management now seems to be the Male Annihilation Technique (MAT), by using the synthetic pheromone for *Bactrocera cucurbitae*. The male population gets reduced very fast, and indirectly the females, and the total development of broods and their total population gets reduced drastically, so that the pest becomes manageable. Collection and destruction of affected fruits, ploughing soil to kill the pupae, bait sprays and use of food lures or other kairomone traps using coconut shell or other small devices are components of IPM. Snake gourd semilooper is an important pest in that crop. Henosepilachna beetles, pumpkin beetles, and serpentine leaf miner are also problems, but can be managed by timely monitoring and later attack by sprays. There are many pests specific to specific cucurbits, and should be dealt with accordingly.

Custard apple is sometimes attacked at the base of the fruit by mealy bugs.

Davana is sometimes attacked by semiloopers and pentatomid bugs.

Fennel is attacked thrips, semilooper and pentatomid bugs.

Fruit midges that cause galls are the major pests in Fig. Occasionally the stem borer Batocera, and leaf eating caterpillars become problematic.

French basil or ram thulasi is attacked mostly by the lace wing bug, but sometimes also by aphids.

Ginger and turmeric have common pests, and by far, the shoot or stem borer, conogethes is the most important pest in the vegetative stage. During maturation of the underground stem, there is usually attack by many species of rhizome maggots, and usually is combined with a fungal or bacterial rot. The rhizomes are to be uprooted and observed and preventive action taken immediately on noticing the first attack itself. Leaf rollers and leaf beetles are occasional pests only.

Gladiolus thrips attack causes reduction in yield and marketability of the flowers and should be dealt with immediately. Cutworms cause the damage unseen, usually are night feeders or are not easily seen.

The major pests in grapes are grapevine stem girdler, *Sthenias grisator*,grape vine flea beetle, *Scelodonta strigicollis;*grapevine thrips, *Rhipiphorothrips cruentatusand* the grapevine mealy bug,*Maconellicosus hirsutus*. Grapes, being consumed raw, allows only minimal use of insecticides. As a general rule, even when insecticide use is warranted, use botanicals or chemicals with low residual and mammalian toxicity. Monitoring and managing at initial pest build up is the rule. Biocontrol of thrips, aphids and mealy bugs should be promoted. Monitoring of these in the blooming stage and immediate remedies based on the surveillance reduces pesticide application. Release *Cryptolaemus montrouzieri* as a predator against mealy bugs and scales. In severe infestation, use Dichlorvos + fish oil rosin soap. Against stem girdler, deterrent chemicals/botanicals like neem oil/NSKE/garlic extract can be sprayed on the stem during the incidence season. Remove and destroy loose bark and plant parts after every pruning.

Guava has many pests and the important ones are the spiralling whitefly: *Aleurodicus dispersus,* Guava scale, *Chloropulvinaria(pulvinaria) psidii;* Tea mosquito bug, *Helopeltis antonii* , Miridae and the fruit borer, *Conogethes (Dichocrocis) punctiferalis.*

Isabgol grows as a wild grass, and aphids are sometimes pests on the stems.

The next crop is jackfruit which is relatively pest free. But one unseen or unrecognised pest is the fruit or bud borer on the female flowers. This is a complex, involving at least three species, all belonging to Glyphodes.

(Diaphanea). They are the shoot and fruit borer, *Diaphania (Margaronia) caesalis;*leaf caterpillars,*Glyphodes sibillalis* and *G. bivitralis.* There is also the bud weevil/leaf-eating weevils, *Ochyromera artocarpi, Onychocnemis careyae, Telurops ballardi.* The presence can easily be noted during the bud emergence and can be promptly solved by insecticide application, since the crop is assuming importance now, as a food security crop and a major fruit in summer. Monitoring the population of mealy bugs and whiteflies is to be done regularly. Plough the soil before every flowering period to expose pupae of fruit flies. Remove the fallen fruits and bury them deeply with lindane 1.3 D dusted over them. Release Cryptolaemus beetles @ 10/tree during the beginning of the season to prevent mealy bugs and whiteflies. Spray neem formulations@ 2ml/l at monthly/bimonthly interval at the time of flowering, if need be. Under heavy population, use Triazophos 2ml+ neemoil 5ml/l or phosalone 1.5 ml + neem oil 5ml/l. Against tea mosquito outbreaks,(which is almost never there) use malathion 50EC 2ml/l, endosulfan 35EC 2ml/l or fenthion 80EC 1ml/l.

Jasmine is a very important flower crop. There are major pests like the Budworm, *Hendecasis duplifascialis* ;whose attack causes sizeable flower loss by making a hole on the buds, and feeding inside, leaving behind faecal pellets entangled in silk strands. There is also the leaf webworm,*Nausinoe geometralis* whose larva young larvae scrape the tissue, and mature larvae web the leaves together with loose silken threads, which cause burnt appearance due to leaf drying. The blossom midge *Contarinia maculipennis* causes the bud to show purple discolouration 4-5 days after oviposition. Another pest, a non insect, is the *eriophyid* gall mite *Aceria jasmini* that feeds on the leaves and flowers and cause malformation and formation of felt outgrowths in vegetative and floral parts and consequent reduction in growth and yield. There is also the flower thrips.

There is the large cardamom or amomum, that is attacked by the capsule borers and thrips.

The aromatic lemongrass, yielding the lemongrass oil being in the grass family and used for oil extraction alone, is hence to be viewed that way only. Shoot borers, aphids and spittle bugs are sometimes pests, but never considered to be serious.

Litchi is attacked by the fruit borers *Criptophlebia illepida, Rapala varuna, Deudorix epijarbas, D. isocrates,* all in the lycaenidae butterfly family .The dipteran fruit fly, *Bactrocera dorsalis* is the other pest of importance. It is also attacked by the eriophyid mite, *Aceria litchi* and by the bark-eating caterpillars, *Indarbela tetraonis; I. quadrinotata*. These pests have to be monitored systematically and management taken. Loquat also has the fruit fly and Indarbela, and also the flower thrips that may be important.

In mango, the most important pest is the dipteran fruit fly, *Bactrocera dorsalis*. This is the pest that spoils the ripening fruits and is the "worms" in mango, as told by the common man. The management has to start with ploughing of the base of the plant to expose the pupa, collecting and destroying the rotten fruits, applying bait sprays and food lure traps, using the male annihilation technique(MAT), and also the hot brining or the hot salt water technique. The mango hoppers have to be monitored, and at the beginning of the increased population above the visual ETL, managed by insecticides. Mango stone weevil and mango stem borers cause damage in some areas.

Marigold is a poor man's' flower crop that is grown extensively in Tamil Nadu. It has the bud caterpillars, *Helicoverpa armigera;* aphids, *Aphis gossypii* and the thrips *Neohydatothrips samayunkur*.

Mentha is attacked by the cabbage semilooper, but most attack is by the leaf webber and helicoverpa.

Mushroom is now commercially grown for vegetable and is troubled by the sciarid flies. There are also phorid flies that are dung flies and the mushroom mite, *Tyrophagus* that causes heavy damage.

Nutmeg has very good price now a days but is insect free except the jumping lice and scales.

The introduced crop oil palm is now gaining acreage in the southern states. It has the same pests as of coconut, the red palm weevil, the rhinoceros beetle and the coreid bug. The pollinator weevil *Elaeidobius kamerunicus* should not be confused as a pest and should be protected for better yields.

Olive is relatively less attacked by pests, except the scales.

Pests of onion and garlic are dealt together. *Thrips tabaci* may be important and cause curled tips and shining silver streaks. Maggots of the Onion fly,*Delia antiqua* feed on the developing bulb underground and cause rotting. Plants wither, and attack predisposes to infection by fungus causing soft rot of onion. Sometimes cutworms are pests.

The drug crop opium is next. Cutworms, of the species Agrotis usually cut young plants at ground level. Helicoverpa is also troublesome.

In orchids, the orchid scale *Diaspis* cause yellowing and drying of plants, and the same type of damage is done by mealy bugs and thrips also.

Papaya is attacked by the fruit flies, *Bactrocera cucurbitae* and *Toxotrypana curvicauda,* Tephritidae, but recently the papaya mealy bug: *Paracoccus marginatus,*Pseudococcidae has become a problem. In newer areas, it may be a problem, but can be effectively managed by the parasitoids *Acerophagus papayae Anagyrus loecki* and *Pseudleptomastix mexicana .A. papayae* is most effective in the southern states, and has almost fully solved the epidemic attack in a matter of just one year, adding one more classical success ofbiological control in the history of Indian applied bio control.

Patchouli, the medicinal plant, is attacked by leafwebbers, leafrollers and pentatomid bugs, but in all these small revenue crops, monitoring, and mechanical control is only feasible and economical.

Peach is attacked by the green peach aphid/peach leaf curl aphid: *Brachycaudus helichrysi,* Aphididae and also by the peach fruit flies, *Bactrocera zonatus* and the oriental fruit fly, *B. (Dacus) dorsalis.*

Pepper is the king of spices for which the Portuguese came to India through the route discovered by Vasco da Gama. Its most important pest is the minute pollu beetle, which requires systematic IPM in places of epidemic attack. Usually, the crop has less than ETL attack, and hence is ignored. The top shoot borer is serious in plantations, but once the plantation becomes older, can easily be ignored. Root mealy bugs have become a recent problem that sometimes warrants immediate chemical control. There are also the root grubs, in lose soils that eat the roots, but can be managed by IPM.

Periwinkle is a minor crop that has few pests.

Pineapple is relatively free of insects, except the mealy bugs,*Dysmicoccus brevipus;* and the termite, *Odontotermes* spp. Both are serious when they attack the roots. Mealy bug causes shrinking of the leaves, pinkish colouration and arresting of growth. In addition to chemical control, application of the fungus, *Verticillium lecani* on the crown and allowing it to drip down the plant, has shown very promising results.

In the case of plum, the pests are, Plum curculio, *Conotrachelus nenuphar* and other defoliator weevils like *Amblyrrhinus poricollis, Myllocerus laetivirens, M. sabulosus and M. undecimpustulatus*. The curculio is a pest in the grub stage that feeds on the fruits. The San Jose Scale: *Diaspidiotus(=Quadraspidiotus) perniciosus,* is however, very important and the next pest of any significance can be the larvae of the Leopard moth,*Zeuzera multistrigata* which are stem and root borers .

The pests of pomegranate include fruit borer (Anar butterfly), *Deudorix (Virachola) isocrates* and *D. epijarbas* and a very close species, *Rapala varuna*. The nymphs of the mealy bugs *Planococcus lilacinus, Planococcus citri* and *Ferrisia virgata,* infest flower buds and fruits in large numbers and desap. Severe infestation leads to dropping of flower buds and low fruit-set. The fruit becomes unmarketable.

Rose, the sweet flower is usually killed by the attack of hard scales and the black fly. The flower has the major problem of rose thrips and in heavy attack systematic insecticide spray becomes necessary at least for a few weeks. The leaf cutting bees, bud borers and chaffer beetles also are important.

Sarpagandha is a medicinal plant and has some minor pests.

Strawberry and walnut has the walnut weevil and is also prone to cutworms, white grubs and hairy caterpillars.

Our next crop of significance is the sweet tuber, sweet potato, but is attacked by the sweet potato weevil, *Cylas formicarius,* that causes the tuber to be unfit for consumption. There are pheromone traps for the pest, which are quite effective. There are many sweet potato beetles, that often make their entry into question papers, along with sphingids and mealy bugs.

Pests of tea are the flush worms that cause the maximum damage. Tea mosquito is also equally important, and another non insect pest of importance is the tea mites.

In vegetables, one important crop is tomato. One of the most important pests in tomato is the fruit borer, *Helicoverpa armigera.* Young larva feeds on tender foliage and starts infesting the fruits from later instars. They make circular holes and thrust only the head and a part of its body into the fruit and eat the inner contents. The rest of the body is on the outside. A single caterpillar can destroy many fruits

The larvae are cannibalistic and can feed on one another when food is scarce, but almost never occur in the field, but is most common in lab rearing and hence warrants individual isolated rearing in semi synthetic diet. Collect and destroy the infected fruits and grown up larvae. IPM involves collection and destruction of infested fruits along with the larvae, trap cropping, pheromone trapping,release of parasitoids, and predators, spraying of NPV, and lastly, use of minimal quantity of specific insecticides.

Tomato is also highly susceptible to the serpentine leaf miner, *Liriomyza trifolii* , which got introduced into India in the last decade due to deficient quarantine. Their maggots mine into leaves and causes serpentine mines, and later drooping and drying of leaves. Leaf miner is a major problem in all tomato growing areas both in the open and in green houses. The maggot just after hatching makes narrow, linear mines in the leaf by feeding the mesophyll tissues. In severe cases leaves show extensive mining which appear as zigzag silvery lines on the upper surface of leaves. The infested leaves turn yellowish, then browning and later dry up.

For management, it is important to monitor the pest occurrence regularly. On sighting the first of the leaf mines, they have to be systematically collected and destroyed by thrashing the leaves. Apply NSKE 3% or other neem formulations, by dusk or at least very early in the morning when the maggots cone out of the mined leaves for getting into fresh areas. When severe, remove the affected leaves, destroy them and spray dichlorvos, malathion or other insecticides that have less than three days of residual action. The leaf caterpillar,*Spodoptera litura,* whitefly *Bemisia tabaci,* thrips in summer, and epilachna beetles cause damage occasionally.

The last crop in the book is the flower crop, tuberose, which is attacked by the thrips, on its flower, the two spotted spider mite, and some times, by helicoverpa also.

An additional chapter is the IPM in vegetables. The general strategies have been described here. Reducing insecticide application to the lowest level is the motto of IPM in vegetables. Monitoring for the initial build up and timely mechanical removal of the first population is a must here. Always remove the first affected plants, for diseases like mosaic, and destroy them by burying or burning. Remove insect eggs, insect affected plant parts and adults and immature stages of insects, and destroy them every day. Do not throw them in the field. They will either come back, or pupate and attack later. During nursery stage, use wood ash, cow dung solution, and plant extracts to deter all pests. Use only low residual insecticides like Dichlorvos 1ml/l, Malathion 2ml/l, Endosulfan 2ml/l or Quinalphos 2ml/l, in spots, at the beginning of attack itself. Use pheromone traps against fruit and shoot borer in brinjal, bhindi, diamond back moth in cabbage and cauliflower etc.Set up sticky yellow traps against aphids and whiteflies in cowpea, brinjal, bhindi, cabbage and cauliflower. Where ever feasible, grow trap crops like mustard in cabbage fields for monitoring, or acting as a barrier or also for attract and kill strategy. Always grow resistant varieties and treat the seeds with seed treating dusts (STD) such as imidachloprid, before planting. Release parasitoids wherever available, like *Chelonus blackburni* @ 4000/acre in bhendi, *Trichogramma chilonis* @ 20000/ acre/ week against Helicoverpa and *Cotesia plutellae* against Diamond Back Moth. Predators for the pests have also to be used extensively.

The reader, whosoever he/she may be, has the full liberty to point things out and the gesture will be greatly appreciated. Please feel free to browse the website www.utopianres.com and to send the comments on the book to the e-mail address *ranjith.am@gmail.com or yourfoundation@gmail.com.* Your comments are our strength.

SECTION-I: FRUITS

Chapter - 1

ALMOND

Green peach aphid/leaf curl aphid : *Brachycaudus helichrysi* : Aphididae

These are green aphids that suck the sap from growing buds. Leaf buds become weak and result in poor setting and premature fruit fall.

The pest has a worldwide distribution. In India it is seen in the hills of North India and in nilgiris.

It is a serious pest of peaches, plum, apricot, and almond. The aphids attack the tender portions like new buds, blossoms, petioles, tender fruits and leaves. When the new flushes are attacked, the leaves get curled and crinkled and fruit set is either reduced or fruits fall off. The infestation starts with the commencement of bud swelling and continues until after flowering. The leaf curl may be confused with that due to the fungal leaf curl by *Taphrina deformans*.

Biology

The eggs are laid in the axils and bases of leaves which over winter. The nymphs hatch out 3 to 4 weeks before flowering and when flowering commences, the aphid population reaches around ten per bud. The pest can survive on an alternate host *Erigeron canadensis*. Reproduction is normally parthenogenetic asexual, but one generation of sexual cycle is also made every year. The egg is 0.6 by 0.2 mm and is laid singly. It also infests quince, Chrysanthemum and Gerbera

Management

In Endemic areas, chemical pest management is warranted. A pre bloom spray of methyl demeton, 0.05% or dimethoate 0.05% about a week before flowering will manage the pest. If reinfestation occurs, a second spray 7-10 days after petal fall may be given.

The aphids : *Hyalopterus amygdali* (Blanchard), *Brachycaudus amygdalinus, Myzus persicae, Pterochlorus persicae* : Aphididae

These are also of economic importance, in relation to the high infestation levels that they can reach during spring. The nymphs and adults desap from the shoots causing devitalisation and crinkling of young shoots.

Metallic wood borer: *Sphenoptera lafertei, Capnodes tenebrionis* : Buprestidae

These are really noxious and their larvae can rapidly suppress the infested almond trees; the larvae are wood borers and their chemical control is also very difficult. Surveillance for beginning of stem infestation, swabbing the trees with insecticides and application of aluminium phosphide tablets may be tried to save attacked trees.

Almond weevils : *Myllocerus laetivirens, M. undecimpustulatus, M. discolor, Amylyrrhinus poricollis* : Curculionidae

The adults and grubs cause damage. Young larvae (grubs) feed on the roots of plants. The weevils congregate on the ventral surface of the leaves, nibble irregular holes and gradually eat away the entire leaf lamina leaving only the midribs. The tender leaves are eaten first and then the older leaves are also damaged. The adult weevils cut holes on the leaf or eat the margins of leaves. The grubs feed the roots of plants and cause wilting and drying of plants. But these weevils are generally of minor importance, due to low populations of both the grubs and adults, and do not require specific chemical management.

Biology

Adult weevils are small (3-4 mm) with pale metallic green or grey or brown colour. Adults appear in May and lay eggs in soil in groups of 40 -50/ batch and hatch in 4-5 days. They lay about 300-500 eggs in the soil over a period of three months. The grubs hatch out in a week and start feeding on the roots of the hosts. The grub period is around 10 months. The full grown grubs pupate in earthen cells for 10-12 days. The emerging adults take to the aerial portions and start feeding on the leaves of the hosts. They are polyphagous and can feed on a variety of crop plants and other vegetation.

Management

In cases of considerable damage to the roots in a plantation, the root zone can be drenched with 0.1 % Chlorpyriphos or rake up a dust formulation of Chlorpyriphos or Lindane 2% dust or any other dust formulation in the soil. The adults can be warded off or killed by application of contact insecticides like chlorpyriphos 0.05%, Malathion 0.05% or Quinalphos 0.05% or synthetic pyrethroids at 0.01-0.05% depending on the chemical nature.

Tent Caterpillar: *Malacosoma disstria*: Lasiocampidae

Defoliation caused by tent caterpillars may be serious on young orchards. They are voracious leaf feeders, and settles in the centre of the twig while at rest. Young caterpillars skeletonize the leaves, whereas older caterpillars consume the leaf, leaving only the midribs. On small trees, infested twigs may be cut out and destroyed. This being a minor pest, is usually taken care of by sprays for other pests. Spray programs for other insects generally reduce populations. If insecticide treatments are required, localized treatments on individual trees and branches are sufficient. Treat when small caterpillars are first observed. Popu-lations usually decline naturally due to disease.

Apricot scale/European Fruit Scale : *Lecanium Arthenolecanium corni*: Coccidae

These are hard scales that drain the sap from the stem, and cause an unhealthy look for the plants. Scales are very difficult to remove, except during their first instars, when they are soft, and move for a resting site. At such times, when the crawlers emerge, give a spray with any contact insecticide, with liquid soap added to the solution @5ml per l of spray volume.

Lecanium scales

San Jose Scale: *Diaspidiotus (=Quadraspidiotus) perniciosus* Diaspididae

The San Jose scale derives its popular name from San Jose, California where Comstock discovered and named it in 1881. It has been considered the most pernicious (injurious, destructive) scale insect in the United States. The adult female is round, slightly convex with a black pustule. The adult males are small and winged. The body of adult female is yellow and is covered with a rounded dark gray scale up to two mm in diameter. Eggs are not seen because the females give birth to young ones. Yellow crawlers are born viviparously and emerge from the inside at the rate of two or three a day. The crawlers then move out to disperse to other parts of the plant, find out new feeding sites and start feeding. They moult after about ten days and begin to lose their eyes, legs and antennae. The adult female appears after the next moult and the scale develops, incorporating the larval exuviae. The development of the male involves three moults. The male nymph is more elongate than the female and the adult male is orange coloured and has wings. It lives only for a few hours.

The nymphs and adults of this insect congregate in large numbers on the bark and desap the trees. Later infestation may spread to the new flushes, blossoms and fruits. These parts are also shrunken, blemished and damaged due to the desapping. The infested region in bark becomes reddish pink and there is purple colouration on the fruits.

Over wintering nymphs become active in mid March and the males emerge in April. Females are ovo-viviparous and reproduce in mid May. About 200-400 nymphs are released by each female, in a month. Crawler period is about 20 days. Total life cycle occupies 35-40 days. 4-5 generations are seen in a year before they start hibernation in mid October.

Management involves use of oil sprays and biological control. Pheromone traps are used for monitoring of the pest. Use dormant oil during the dormant period. Insecticide Dimethoate 0.05%, chlorpyriphos 0.05% or any newer molecules active against sucking pests can be used.

Almond beetles/Chrysomelids : *Mimestra cyanura, Altica* sp., *Haplosoma* sp., *Merista* sp., *Monolepta* sp., *Phyllotreta* sp., Chrysomelidae

These chrysomelids are small coloured beetles with black spots and other markings on the elytra. Adult beetles of all the species appear in swarms during May, defoliate the trees and cause consequential yield loss.

Chapter - 2

AONLA or Indian Gooseberry Amla or Amalalsa

This crop is gaining importance as a main ingredient in Indian diet due to its medicinal properties. Large areas are under its cultivation and the larger fruited varieties have become very popular. It is one of the prime constituents in ayurvedic preparations

Bark-eating caterpillars - *Indarbela quadrinotata, I. tertraonis:* Metarbelidae

The larvae of this species make zigzag galleries and feed on the bark under silken ribbon- shaped webs. They are stout and dirty brown in colour.

Fruit- borer : *Deudorix Isocrates* : Lycaenidae: The larvae of this beautiful butterfly cause considerable damage (up to 42%) to fruits. They start feeding while the fruit is tender, and growing. During severe infestations, chemical control may be warranted. This can be done by spraying contact insecticides like malathion, or dichlorvos or fenvalerate or other synthetic pyrethroids.

Shoot-gall maker: *Betuosa stylophora* : Thyrididae

The young caterpillars of these bore into tender shoots and cause gall formation.

The gall resembles "snake chamber's flute" at the apices of growing twigs. The infestation adversely affected the growth of twigs. All varieties are susceptible. Growth of severely infested shoots is suppressed .This will prevent the formation of newer shoots; suppress the shoot and flower production, and consequently, the yield.

Gooseberry gallmite: *Eriophyes grossulariae*: Eriophyidae

They make galls on the plant

Gooseberry gall midge/ bud deformer: *Rhopalomyia grossularis:* Cecidomyidae

This pest makes the young buds to forn into a rosette

Aphid*:* *Ceriaphis (=Schoutedenia?) emblica* Aphididae

The nymphs and adults suck the sap from leaves and new flushes. They will also feed on the flowers when they bloom

Leaf roller : *Gracillaria acidula* : Gracillariidae

Small yellowish-white tiny larvae of leaf roller initially mine the leaflets but advanced instar larvae roll the leaflets. They feed on the chlorophyll content by scrapping the rolled leaves. The damaged leaflets fall down and only midrib remain on the twig.

Bark and shoot borer : *Indarbela tetraonis* : Metarbelidae

The larvae of these bark borers can feed on the bark and cause devitalisation of the tree.They are usually seen at the base of the tree.

Mealy bug : *Nipaecoccus vastator* : Pseudococcidae:

The nymphs and adults infest the leaves and tender shoots and cause devitalisation due to sucking of cell sap.

Aonla hairy caterpillar : *Selepa celtis* : Noctuidae

These feed on the leaves and young shoots.

Chapter - 3

APPLE

Apple codling moth: *Cydia (Carpocapsa) pomonella*: Tortricidae

This is an international pest and is now seen in Kashmir, Himachal Pradesh and Tamil Nadu. This is a fruit borer whose moth is silvery white with a dark band on the margin of the forewing and light brownish hind wings with a fringed margin. The moths hold their wings roof-like over the body while at rest.

Symptoms

Female lays white coloured, flattened eggs, singly on developing fruits, leaves and twigs. Egg period is 4-12 days and larval period occupies 21-30 days inside the fruit. Newly hatched larvae searches for the fruits and enter either at the calyx end or through the side. They bore through the skin and feed on the fruit flesh for a few days, then move towards the core where they feed on seeds and flesh. As they feed, they push excrement out of the apple through the entry hole, which is gradually enlarged and often serves as an exit hole. When the larvae grow up, they come out of the fruit and falls on ground and reach the bark of the tree for shelter in cracks and crevices to construct a silken cocoon. The pupal period lasts 8-14 days. It is an international pest for over 200 years .

Management

Codling moth can be very difficult to manage, and moth numbers should be kept low from the beginning of the season.

Use of resistant varieties can be done while planning orchards. Thin the crop by removing the affected fruits and destroy them. Monitor for the first fruit damage and always remove and destroy them. Sticky pheromone traps to attract the males are also available. They could be used for monitoring and also to reduce the population to an extent by the male annihilation technique(MAT). There is also a Codling moth granulosis virus. Chemical control is by using Spinosad and carbary.

San Jose Scale: *Diaspidiotus(=Quadraspidiotus) perniciosus:* Diaspididae:

The San Jose scale derives its popular name from San Jose, California where Comstock discovered and named it in 1881. It has been considered the most pernicious (injurious, destructive) scale insect in the United States. The adult female is round, slightly convex with a black pustule. The adult males are small and winged. The body of adult female is yellow and is covered with a rounded dark gray scale up to two mm in diameter. Eggs are not seen because the females give birth to young ones. Yellow crawlers are born viviparously and emerge from the inside at the rate of two or three a day.

The crawlers then move out to disperse to other parts of the plant, find out new feeding sites and start feeding. They moult after about ten days and begin to lose their eyes, legs and antennae. The adult female appears after the next moult and the scale develops, incorporating the larval exuviae. The development of the male involves three moults. The male nymph is more elongate than the female and the adult male is orange coloured and has wings. It lives only for a few hours.

The nymphs and adults of this insect congregate in large numbers on the bark and desap the trees. Later infestation may spread to the new flushes, blossoms and fruits. These parts are also shrunken, blemished and damaged due to the desapping. The infested region in bark becomes reddish pink and when the fruits are attacked, they have purple colouration.

Over wintering nymphs become active in mid March and the males emerge in April. Females are ovo-viviparous and reproduce in mid May. About 200-400 nymphs are released by each female, in a month. Crawler period is about 20 days. Total life cycle occupies 35-40 days. 4-5 generations are seen in a year before they start hibernation in mid October.

Management involves use of oil sprays and biological control. Pheromone traps are used for monitoring of the pest. Use dormant oil during the dormant period. Insecticide Dimethoate 0.05%,chlorpyriphos 0.05% or any newer molecules active against sucking pests can be used.

Woolly apple aphid: *Eriosoma lanigerum*:Pemphigidae

Woolly apple aphid is a sucking insect pest that attacks the shoots, branches and also the roots of the tree. The colony of these aphids appear as a cottony mass generally clustered in wounds and pruning scars on the trunk and branches of the tree. The aphids are purplish in color surrounded by white, cottony, thread-like secretions. It gets its name from the woolly appearance of its colonies. Long strands of white wax are produced that help to protect the colony of white fluffy aphids from predators and pesticide sprays. They also have the winged forms or alates.

Reproduction is both sexual and asexual but mostly by parthenogenesis. Aphid is active throughout the year except in cold months of December and January. Wingless forms are present all through out the year whereas winged forms are seen from July to October. Winged form disperses by flight and gives rise to apterous forms by sexual reproduction. In winter, colonies on the tree trunks move to the base of the trunk just below the soil. About 110 young ones can be produced by each female. Nymphal period can last for 35-42 days with four instars.

The nymphs and adults of these aphids live in colonies, clustered on the roots or branches of apple, pear, elm etc. On the aerial parts, it is seen as a woolly mass. Damage is caused by sucking of sap from aerial parts and roots, resulting in gall formation, stunting, and retardation of growth and proliferation of fibrous roots. They also cause weakening and death of the smaller plants, galls on the roots and white woolly patches on the trunk.

Management involves use of resistant root stocks, and the parasitoid, *Aphelinus mali*. Use dormant oil during the dormant period. Insecticides like Dimethoate 0.05%, Chlorpyriphos 0.05% or any newer molecules active against sucking pests can be used.

Cottony cushion scale : *Icerya purchasi* : Margarodidae : Hemiptera

This insect is the central pest in the classical biological control that the world has ever seen. The pest had invaded the apple trees of California and caused a near disaster. It was by the use of the coccinellid predator *Rodolia cardinalis*, that the insect was finally controlled. This has, since then become history's most successful celebrated cases of classical biological control. The female is with a cottony ovisac. Nymph is a pinkish crawler with long antenna and group of hairs.

The desapping by nymphs and adults cause yellowing of leaves. In severe instances, the trees become unhealthy, flowers and fruits are not formed, and

there is heavy yield loss. The pest is classically controlled by the vedalia beetle, *R.cardinalis*.

Blossom thrips : *Thrips rhopalantennalis*, T. *flavus*, *T. florum*, *Haplothrips tenuipennis* : Thripidae

These are tiny, yellow, pale-brown and black slender insects seen within the flowers and flower buds. Adults and nymphs lacerate floral parts and suck sap resulting in brown spots. Infested flowers give sickly and faded look. Lesions cause flower fall, inferior fruits and loss in yield. Attack on tender leaves causes necrotic silvering, curling, deformation and withering.

Defoliating and fruit eating beetles/Cockchafers or May-June beetles: *Adoretus duvauceli, A. versutus,Anomala dimidiata, A. flavipes, A. lineatopennis, A. rufiventris, Holotrichia longipennis, Hilyotrogus holosericeous, Lachnosterna coriacea, L. crinicollis, L. longipennis, Mimela passerine, M. pectoralis.*

Green melolonthid: *Anomala dimidiata:* Melolonthidae

These are pests in adult stage, where the different coloured beetles (Reddish brown, blue-black, and green), feed on foliage and developing fruits. The adults appear in May- June and lay eggs in soil at dusk during rainy season. They attack all temperate fruit plants whose leaves get skeletonized. The small fruits are eaten and fall prematurely. Their immature stages (white grub) feed on almost any root; crops, weeds, potato, vegetables and grasses.

Apple tortrix/Leaf- roller and fruit- scrapper: *Archips fuscocuprianus*: Tortricidae

The black- headed green caterpillars of this pest fold leaves of apple from April-May onwards. Attack is seen on foliage during summer and last generation of caterpillars attack fruits in field and also in storage and damages them by scrapping. Sometimes cent per cent damage is done to the stored apples.

Leopard moth: *Zeuzera multistrigata*: Cossidae

The adults are white moths with attractive patterns, seen at dusk during May-June. The eggs are laid singly in cracks of barks. Within about a month, the emerging pinkish- white young caterpillars bore into branches and stems. The early instars scrap off the cambium and induce gummosis and later instars enter into the hardwood and make tunnels up to 25cm in the terminal shoots, secondary branches and main stem. This can result in drying of shoots within 1-2 years and of the whole plant within 3-4 years.

Apple fruit moth: *Argyresthia conjugella:* Argyresthidae: Lepidoptera

These are small moths whose larvae are greyish in colour. The larvae bore into the fruits and feed on seed. The infested fruits show exit holes. Fully-grown larvae emerge out after tunneling through the pulp of the fruit and render it unfit for use.

Apple zygaenid : *Aglaope phylina* : Zygaenidae

Moths are white in colour and the slug – like larvae mimic plant colour. Caterpillars feed on the bark, leaf and flower spurs resulting in defoliation, reduced flowering and fewer yields.

Apple Root borer: *Dorysthenes hugelli* : Cerambycidae

The chestnut red beetles lay eggs in soil during July- August. The emerging grubs feed on thick roots and other organic matter for long periods. Their longevity is around 3.5 years and they a have preference for sandy soil. Root borer attack makes the plants weak and shaky and can cause death of plants.

Tent caterpillar : *Malacosoma indica* : Lasiocampidae

Brown coloured moths lay eggs in bands around small twigs in May. The larvae are dark brown to black, 40-50mm in length. These larvae spin triangular or quadrangular tent of silk at forking of twigs and remain there during day. Caterpillars feed gregariously on leaves at night, tracing their path through silken threads.

Apple hawk moth: *Langia zenzeroides :* Sphingidae

These are large moths. The huge larvae feed on large quantity of leaves.

Hairy caterpillars: *Euproctis signata:* Lymantridae

These occur sporadically; caterpillars are profusely hairy. They feed voraciously and defoliate trees.

Indian gypsy moth: *Lymantria obfuscata:*Lymantridae

Eggs are laid by the bulky adult female moth, which has only rudimentary wings and hence cannot fly. Eggs are laid under the bark on the tree trunks or on fallen leaves. They feed on many shade trees and other trees also. Feeding is gregarious and done during night. The trees are almost completely defoliated

Apple leaf miner: *Gracillaria zachrysa* :Gracillaridae

Young caterpillars make several mines on leaf surface; later they leave mines, roll young leaves longitudinally into tubular or conical pouch and feed within; the maximum damage is done during summer (April-May) and in autumn (September-October).

Apple stem borer: *Apriona cinerea, A. sarta* :Lamiidae

These are ashy-grey to greyish –yellow, long horned beetles with conspicuous blackish tubercles at the base of elytra. The grubs cause damage to stem and shoots as a result of which plants may become weak and may even die.

Attacked branches are with small circular hole and mass of excreta and chewed up wood particles protruding out. The bark is gnawed and leaves defoliated and the shoots show circuitous galleries. The trunk is hollowed out and the infested trees remain stunted.

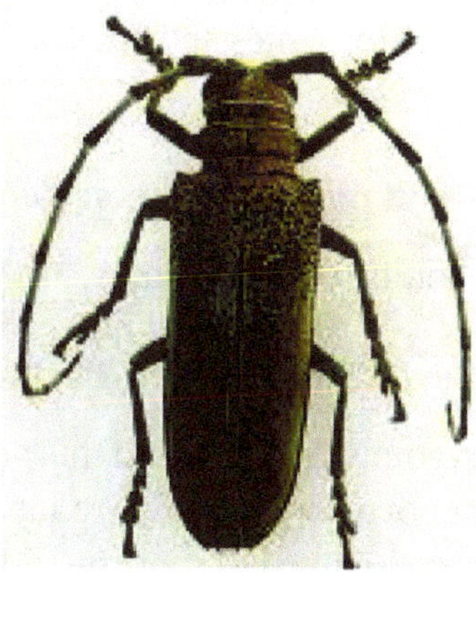

Female lays eggs inside a cavity, which is excavated on shoots. Egg period is for 7-9 days. Grubs undergo hibernation during winter and resumes feeding in March. They pupate inside the tunnel. The pupal period can be for 30-35 days.

Stem borer: *Batocera rufomaculata* :Cerambycidae

These are dark brown beetles that could be above ground borers or root borers. Attack is evidenced by frass and faecal pellets or wilting of leaves and drying of terminal shoots.

Shot hole-borer : *Scolytus amygda, S.faciatus, Scolytoplatypus raja, Pitygenes* sp., *Xyleborus* sp. : Scolytidae

These are small dark coloured beetles with a deflexed head. The grubs of these bore into the trunk. There is an exudation of a dark fluid. The trees become weak and the branches dry up.

Stem borer : *Aeolesthes holosericea* : Cerambycidae

These are serious stem borers in India. The grubs damage the trees and cause even death of the trees. This pest also attacks a lot of forest trees and freshly felled trees, apart from cultivated trees like jack. The eggs are laid on the bark and emerge out into grubs that bore inside and start feeding by making tunnels. The pupation is inside the tunnel and the adults emerge out from inside.

Fruit fly : *Bactrocera zonatus* : Tephritidae

These can infest the fruits occasionally and the maggots feed inside the ripening fruits causing rotting

Fruit piercing moths: *Calpe aphideroides:* Noctuidae

Adults of this pest pierce the fruits and suck the sap from ripe apples

Psyllid:*Psylla mali*: Psyllidae

Nymphs and adults of this pest are seen congregating on the fresh leaves and shoot endings. They are sap feeders and severe sap sucking can also lead to development of sooty mould.

European red mites: *Panonychus ulmi:*Tetranychidae: Acarina
Two- Spotted spider mite: *Tetranychus urticae*

Mites can produce light infestation which result in speckling of the foliage, or heavy infestation when the foliage becomes pale, sickly or bronzed. They feed on the foliage causing loss of chlorophyll, bronzing followed by leaf drop and weakening of buds. This results in poor size and quality of fruits. Excessive feeding results in fruit drop during the season and reduced blossom and fruit set in the subsequent years.

IPM in Apple Orchards

Insect Pest Management in apple orchards involves three main components – monitoring, damage assessment and preventive action. Year round management strategy is based on six season- crop stage periods. These are the dormant winter, spring, blooming, post blooming, summer and pre-winter harvest stages.

Dormant winter: Monitor and assess mite and aphid egg population and San Jose scale. Red mites are best controlled by the application of dormant oil. Allow low populations of rust mites to survive, so that predatory mites feed on them and control red mites later in the season. Normally, natural enemies often control the green aphids. A delayed dormant oil spray will take care of early damage. Only if more than 60% of the terminals of trees are affected, especially in young orchards, a chemical like Imidachloprid 5ml in 2l or Diazinon 3ml/l can be used. Do not repeat the spray with the same insecticide. Dormant oil spray has to be done annually to manage San Jose Scale in orchards where the scale is endemic. There are ladybird beetles which are good predators, but insecticides during the growing season will kill them.

Spring: During spring the aphids will be a problem if IPM was not done earlier. Caterpillar pests like leaf rollers and orange tortrix are mainly seen during this period.

Blooming: Codling moth monitoring and mating disruption with pheromone traps has to be done. Monitoring has to be continued for aphid and caterpillar pests.

Post blooming: Fruit damage by codling moth will occur only if pheromone trapping and mating disruption was ineffective. Monitor also for aphids, leaf rollers and San Jose scale crawlers. If crawlers are abundant, as evidenced from sticky trap catches, a chemical control using chlopyriphos or dimethoate may be needed.

Summer: Trapping and mating disruption controls codling moth. Previous IPM could have controlled leaf roller, tortrix, leaf miners and lygus bugs. Use a contact spray only if the populations are heavy.

Pre-winter harvest stage: Check fruit for codling moth larvae. Remove apple, pear and walnut trees in nearby abandoned orchards to destroy reservoirs of codling moth. Remove affected fruits and harvesting tools and containers from orchard and destroy affected fruits. Collect and destroy dropped fruits regularly.

Chapter - 4

APRICOT

Indian gypsy moth: ***Lymantria obfuscata:*** **Lymantridae**

Eggs are laid by the bulky adult female moth, which has only rudimentary wings and hence cannot fly. Eggs are laid under the bark on the tree trunks or on fallen leaves. They feed on many shade trees and other trees also. Feeding is gregarious and done during night. The trees are almost completely defoliated. Management involves spot application of contact insecticides at the beginning, when the larvae are small and before they disperse to the different leaves.

Apricot chalcid: ***Eurytoma samsonowi:*** **Chalcididae**

The chalcids are one of the few crop pests from hymenoptera. Usually pests in chalcididae are seed chalcids. Eggs of this very small chalcid are laid inside young fruits. The grubs feed on the developing seeds. Fruit growth is stunted and fruits fall off prematurely. Pupation takes place inside the fruit. Seed chalcid management involves spay application before the eggs are laid on the young fruits, and should be done in endemic areas.

Apricot weevil: ***Emperorhinus defoliator:*** **Curculionidae**

Adults defoliate trees during summer.

Ash weevils /Grey weevil: *Myllocerus discolor;M. undecimpustulatus; M. laetivirens:* Curculionidae

The adult weevils cause notching of leaf margins. Grubs feeds on roots resulting in wilting and death of plants.

The adults and grubs cause damage. Young larvae (grubs) feed on the roots of plants and cause wilting and drying of plants.

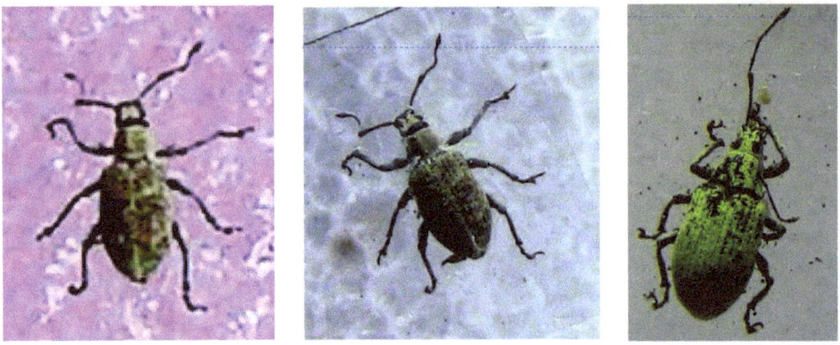

The weevils congregate on the ventral surface of the leaves, nibble irregular holes and gradually eat away the entire leaf lamina leaving only the midribs. The tender leaves are eaten first and then the older leaves are also damaged. The adult weevils cut holes on the leaf or eat the margins of leaves. But these weevils are generally of minor importance, due to low populations of both the grubs and adults, and do not require specific chemical management.

Adult weevils are small (3-4 mm) with pale metallic green or grey or brown colour. Adults appear in May and lay eggs in soil in groups of 40 -50/ batch and hatch in 4-5 days. They- lay about 300-500 eggs in the soil over a period of three months. The grubs hatch out in a week and start feeding on the roots of the hosts. The grub period is around 10 months. The full grown grubs pupate in earthen cells for 10-12 days. The emerging adults take to the aerial portions and start feeding on the leaves of the hosts. They are polyphagous and can feed on a variety of crop plants and other vegetation.

Management

In cases of considerable damage to the roots in a plantation, the root zone can be drenched with 0.1 % Chlorpyriphos or rake up a dust formulation of Chlorpyriphos or Lindane 2% dust or any other dust formulation in the soil. The adults can be warded off or killed by application of contact insecticides like chlorpyriphos 0.05%, Malathion 0.05% or Quinalphos 0.05% or synthetic pyrethroids at 0.01-0.05% depending on the chemical nature.

Chafer beetle : *Anomala polita* : Melolonthidae

Cockchafer beetles that are pests in adult and grub stage. The adults nibble the leaves occasionally. Main damage is due to the grubs of the beetles that eat the feeder froots and cause yellowing and loss of vigour of the plants

Flat-headed borer : *Sphenoptera lafertei* : Buprestidae

Small coppery black adult beetles feed on the foliage. The eggs are laid on the branches facing the sun. The grubs emerge and bore into the stem and then down to the root region. They pupate inside the larval tunnel. The plants may die in severe attack. Periodical monitoring and application of insecticides on the tree stem to ward off egg laying has to be practiced in endemic areas. An occasional pest.

Scale insects: *Eulecanium* sp. Coccidae *Parlatoria oleae* : Coccidae

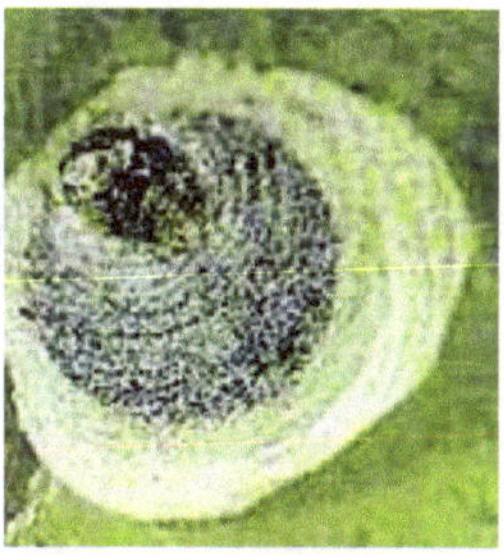

Parlatoria is the olive scale. Both these congregate on the trees and fruit and desap from the parts. Fruit quality and plant quality is affected.

San Jose Scale: *Diaspidiotus (Quadraspidiotus) perniciosus:* Diaspididae

The San Jose scale derives its popular name from San Jose, California where Comstock discovered and named it in 1881. It has been considered the most perni-cious (injurious, destructive) scale insect in the United States. The adult female is round, slightly convex with a black pustule. The adult males are small and winged. The body of adult female is yellow and is covered with a rounded dark gray scale up to two mm in diameter. Eggs are not seen because the females give birth to young ones. Yellow crawlers are born viviparously and emerge from the inside at the rate of two or three a day. The crawlers then move out to disperse to other parts of the plant, find out new feeding sites and start feeding. They moult after about ten days and begin to lose their eyes, legs and antennae. The adult female appears after the next moult and the scale develops, incorporating the larval exuviae. The development of the male involves three moults. The male nymph is more elongate than the female and the adult male is orange coloured and has wings. It lives only for a few hours.

The nymphs and adults of this insect congregate in large numbers on the bark and desap the trees. Later infestation may spread to the new flushes, blossoms and fruits. These parts are also shrunken, blemished and damaged due to the desapping. The infested region in bark becomes reddish pink; purple colouration on fruits.

Over wintering nymphs become active in mid March and the males emerge in April. Females are ovo-viviparous and reproduce in mid May. About 200-400 nymphs are released by each female, in a month. Crawler period is about 20 days. Total life cycle occupies 35-40 days. 4-5 generations are seen in a year before they start hibernation in mid October.

Management involves use of oil sprays and biological control. Pheromone traps are used for monitoring of the pest. Use dormant oil during the dormant period. Insecticide Dimethoate 0.05%,chlorpyriphos 0.05% or any newer molecules active against sucking pests can be used.

Apricot fruit moth: *Cydia funebrana* : Tortricidae

The caterpillars feed on mesocarp of developing fruits and render them useless. The fruits become filled with frass and excreta.

Chapter - 5

BANANA

Pseudostem borer: *Odoiporus longicollis*: Curculionidae

Distribution : It is widely distributed in all banana growing areas and has got introduced into South India somewhere in the eighties and spread fast in the banana areas of Tamil Nadu, Maharashtra, Kerala and Karnataka.

Biology : Adults of the weevil are reddish brown to black and resemble the adults of the rhizome weevil in size and shape. They live for about four months. The snout is long with the characteristic geniculate antennae.

Eggs are thrust within the air chamber in the leaf sheath and pseudo stem through oviposition slits made by rostrum @ one egg/ air chamber. The sites of egg laying may show the exudation of gummy sap. Egg period is for 4-8 days. The apodous grub tunnels inside through the whorls of the leaf sheaths into the pseudo stem and causes the weakening of the plant. The inside can be seen tunnelled and containing the excreta of the grub. The attack usually starts when the initiation of the bunch primordium occurs deep inside the pseudo stem at its soil end. This is about 4-5 months after planting in the case of the nendran

variety. Grubs pupate in a tunnel towards the periphery in a cocoon made from pieces of fibrous materials of sheath. The pupal period may last for about 24-44 days. Due to the attack, the central spindle of the plant is weakened, the bunch does not emerge and the plant gets toppled in the wind. The whole yield is lost.

Management

- Regular monitoring to see the egg laying punctures with ooze of plant sap. Apply insecticides immediately on the pseudo stem sheath.
- Disposal of infested plants by chopping and burning
- Maintaining healthy plantation by periodical removal of dry leaves and suckers
- Spray the pseudo stem with carbaryl @ 4gm/l all around the pseudo stem in endemic areas, once a month, commencing from 5th month. Alternate chemicals are triazophos 1.25ml/l or quinalphos 2.5 ml/l .This has to be continued until flowering.

Rhizome weevil : *Cosmopolites sordidus* : Curculionidae : Coleoptera

This pest is distributed in many countries including India, south East Asia, Australia, the Hawaii Islands, tropical South Africa and tropical America.

Adult weevil is dark brown, stout and measures about 13 x 5mm. Eggs are laid in small burrows, which the beetle scoops out on the rhizome or leaf sheath at the soil level. The emerging grub and later instars are apodous, yellowish

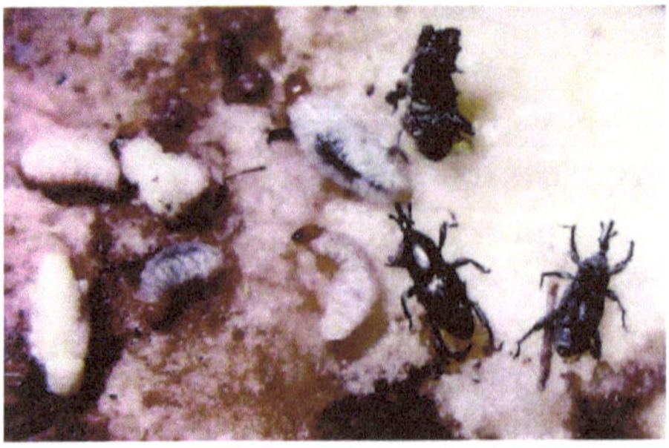

white, with a wrinkled body and red head. These bore into the rhizome causing extensive tunnel-ing. The rhizomes, when split open, will show the tunnels, and are black and weak. The grubs become full-grown in 25 days. Pupation is in a chamber made near the outer surface for 5-6 days. Adults can live for one year.

Due to the attack, the plants show a very unhealthy look, with death of unopened pipe, withering of outer leaves and yellowing. The yield is very low. Bunch weight may be reduced, and sometimes bunches are never produced. The attack is less when short duration varieties like nendran are cultivated. It is more in perennial plantations.

Management

- The best way to prevent attack by the weevil is to go for tissue cultured suckers. This way there is no pseudo stem for the pest to lay eggs or for the grubs to develop. By the time the suckers develop the rhizome, the plant has already grown well. Even then, the rhizome is too small for the weevil to oviposit. However, from the first ratoon of this TC crop onwards, the rhizomes will be susceptible.
- Selecting pest free suckers for planting and renewal of suckers every season with fresh uninfested suckers is the best solution.
- Dip the suckers in insecticide solution containing Carbaryl paste @ 4/100 ml and smear on the pseudo stem. This can also be mixed in cow dung slurry, and the paste smeared on the cleaned rhizome. These are then air dried in shade and planted.
- Trapping the adult weevils by placing chopped pseudo stem in the cropped area is another possibility.
- Soil incorporation of lindane 1.3 D 20 g/plant, carbaryl 5D 10-20 g/plant, carbofuran 3G 10 g/plant or phorate 10 G 5 g/plant around pseudo stem can be recommended in perennial plantations with endemic attack by the weevil

Leaf caterpillar: *Spodoptera litura* : Noctuidae

The adult is a stout grayish brown moth 3.8x1.8 with wavy white markings and criss-cross lines on the forewings and white hind wings having a brown patch along its margin. Young larva scrape leaves on ventral surface, and growing larvae defoliate crops and also feeds on young fruits.

Eggs are laid in masses covered with white hairs on the surface of affected leaves. Eggs hatch in 4-5 days. Larva eats for 14-20 days. They are pale brown with greenish to violet tinge and with stout whitish or creamy bands running the length of the body. Yellow and purplish spots found in sub marginal areas. Faecal pellets are seen on the leaves and on the ground which is the indication of the pest incidence Pupa is seen in the soil for 7-15 days.

Management

In banana, with regular monitoring, this is not a serious pest. However, if the attack is severe, then,

- Set up light trap at 1/ha or pheromone trap at 15/ha with spheroclin SL lure.
- Destroy egg masses and grown up larvae.
- Apply any one of the following insecticides per ha to control early instar larvae (1st to 3rd instar). Carbaryl 10D 25 kg, fenitrothion 750 ml, carbaryl 50 WP 2 kg, quinalphos 25 EC 2ml/lit , fenitrothion 625 ml, phenthoate 1250 ml and dichlorvos 750 ml per ha.
- Spray any one of the following per ha to control the 4th to 6thinstar larvae. Chlorpyriphos 2 lit, dichlorvos 1 lit (76 WSC 2 ml/l), quinalphos 25 EC 2ml/lit.Spray insecticide during evening hrs.

Banana aphid : *Pentalonia nigronervosa* :Aphididae : Hemiptera

Symptoms

It causes indirect damage by transmitting the notorious virus disease bunchy top. The adults and nymphs are seen in colonies on leaf axils and pseudo stem.

Adult aphids are brownish in colour. They have two forms; the winged and wingless forms. Winged or alate forms have black-veined wings. It lives in colonies within leaf-axils, or tender leaves and at base of the culms at the ground level. They reproduce by giving birth to young ones .This phenomenon in aphids is called vivipary. The females live parthenogenetically. 35-50 nymphs may be produced by a female during a life span of 27-37 days. A female gives birth to one nymph each within every 3-4 hours. Nymphs undergo four instars of 8-10 days duration. Total life cycle takes 12-15 days for completion. About 30-40 overlapping genera-tions may be completed in a year.

Management

- At the time of planting, apply 25 g of phorate or carbofuran to the soil. A second application of 12. 5 g in the leaf axil or 25 g in the base at the soil is recommended followed by a third, at 165 days, if attack persists.
- Remove and destroy "bunchy top" affected suckers immediately on sighting in the plantation.

Other pests

Spotted locust: *Aularches miliaris:* Acrididae

This may be important in isolated situations as a defoliator, feeding on young leaves. The pest usually occurs in sporadic forms in banana plantations near forest ecosystems. This is a pest only rarely.

Spittle bug: ***Phymatostetha deschampes*** **:Cercopidae**

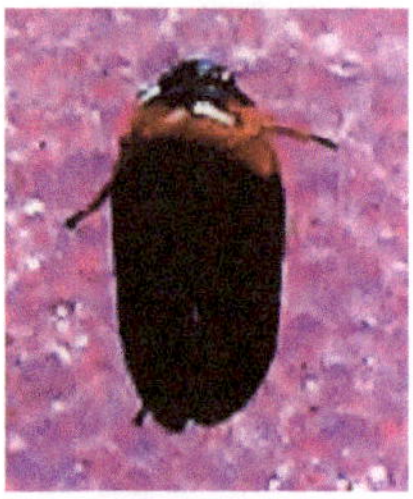

The adults are red and blue spittle bugs, and are seen in large numbers in the rainy season. They suck the sap from tender leaves, congregating on the leaf tips, and pushing out a frothy secretion through the anus. This is only a cosmetic pest.

Fulgorid bug:*Eurybrachys tomentosus* :Fulgoridae

The adult is a large bug measuring about 11 mm and has an olive green tegmina mottled with yellow colour. The abdomen of the female has a mass of white mealy wax. This is an occasional feeder on the leaves and pseudo stem.

Armoured scale: *Chionaspis dilatata* :Diaspididae:

This scale infests the leaves, and occasionally the fruits. Infested fruits do not develop properly.

Scale insect: *Aspidiotus destructor*: Diaspididae: Hemiptera

This is an occasional pest on the leaves causing yellow spots due to desapping.

Soft scales:*Lecanium acutissimum, Vinsopia stellifera*: Coccidae

Occasionally seen feeding on the leaves

Thrips: *Helionothrips kadaliphilus*:Thripidae

These are dark brown in colour. They infest the lower surface of leaves causing blighted appearance and yellowing. In severe cases, the leaves may dry up due to desapping and toxaemia.

Flower Thrips: *Thrips florum, Chaetanaphothrips signipennis*: Thripidae

These may congregate in thousands on the flowers. Due to the desapping, rust like spots may appear on the developing fruits and fruit peels.

Bagworm: ***Acanthopsyche minima, Kophena cuprea*** **:Psychidae**

The males of these bagworms alone are winged. The females are wormiform and enclosed in a case made out of cut pieces of leaves. They eat up holes on the leaf and hang from them.

Nettle grub: ***Latoia lepida:*** **Cochlididae**

These are mostly pests of coconut, but can be seen on banana leaves during rainy season, feeding on the leaves usually during night and remaining within the leaf folds during day.

Slug caterpillar ***Miresa decedens*****: Cochlididae**

The moth is stout and yellowish brown with a velvety golden bloom all over the dorsal side and a pale cross streak on each upper wing. The larva is greenish white with a blackish median dorsal band and four rows of tubercles bearing spines. They are irritating to touch and is a defoliator and pupates in a shell like hard cocoon. Does not warrant control measures except mechanical destruction.

There are other jelly grubs *Belippa laleana, B. Lohor*, and *Microlimax* sp., occasionally feeding on the leaves.

Hairy caterpillar: ***Eupterote geminata*****: Eupteroptidae**

The moth is hairy scaled and yellowish in colour. The caterpillar is also hairy and feeds on the leaves and attains a size of around 6 cm. In the earlier instars, they are gregarious and arranged in curved rows on the leaf. Cocoon is with a dense greyish brown silk and is hairy.

Black hairy caterpillar: ***Pericallia ricini*****: Arctiidae: Lepidoptera**

The adult is a medium sized moth with a wing span of about 5.5 cm. The forewings are dull brown with pale ringed dark spots.

Hind wings are crimson with a marginal and three inner bands of black blotches.

The abdomen also is crimson colored with black bands across. Eggs are laid in clusters and up to 170 eggs can be laid. The caterpillars are gregarious initially, and the last instars are voracious feeders. The body is covered with a dense felt of brownish hairs and pupates on the leaves itself in a cocoon within a layer of hairs. The attack if severe, much of the leaf area, including the slender leaves may be completely eaten. Initial monitoring and destruction is sufficient in most instances. If uncontrollable by mechanical means, use a contact insecticide spray.

Hairy caterpillar: *Asura ila and Creatonotus gangis*

Arctiidae are other hairy caterpillars, that are defoliators on banana.

Spindle fly :*Assuania* sp.: Chloropidae

This is a very occasional pest .The adult fly lays eggs on the spindle surface. The maggots mine the leaves and the slimy mines cause the spindle to get stuck and would not open. A spray of quinalphos or fenthion would control the infestation.

Lacewing bugs: *Stephanitis typicus*: Tingidae

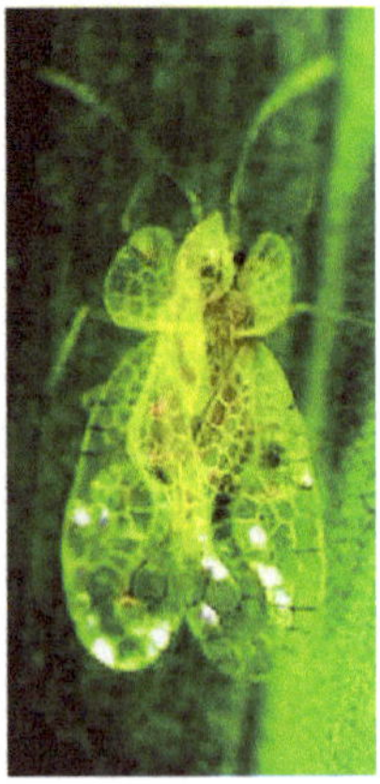

Adults are small, dull-coloured or whitish bugs with lace-like reticulate wings. The nymphs are black coloured. Both nymphs and adults are seen in colonies on undersurface of leaves feeding on sap and cause discolouration of leaves. These are usually pests in the summer months. Usually no chemicals are recommended. If needed, a spray with less residual chemicals can be given. A very minor pests.

Chapter - 6

BER

Ber fruit fly: *Carpomyia vesuviana* :Tephritidae

Adult is a small fly with black spots on the thorax and dark spots on the wings. They have only one pair of wings typical of diptera. The attack by the insect causes rotting of fruits and fruit drop. The adult female lays eggs singly in cavities made on the fruit by the ovipositor. One female can lay up to 22 eggs. Egg period lasts for 2-3 days and the emerging maggots feed inside the fruit for about 7-10 days. The fully grown maggot falls to the ground to pupate in soil. Pupal period is around 5-30 days. There can be 2-3 generations /year.

Fruit Flies : *Bactrocera zonata* : Tephritidae

The adults lay eggs on the fruit. The maggots in the infested fruits cause browning, rotting and fruit drop. In north India, up to 80% fruits are damaged. Thin skinned, soft and sweet cultivars are more susceptible.

Ber fruit borer : *Meridarches scyrodes* : Carposinidae

The adult is a small dark brown moth. Eggs are laid on flowers and tender fruits. The fruit is rendered unfit for consumption. This is a serious pest in south India. Eggs are laid on fruits at pea stage. On hatching, the reddish brown larvae bore into fruits where the rind is thin and soft and feed on the pulp near seed. The extent of damage may vary from 20 to 80 percent.

Leaf butterfly: *Tarucus (Theophrastus) indica* : Lycaenidae

The larvae of this species cause defoliation .They are spindle shaped, and fleshy green. The adult is a blue butterfly with several white spots

Hairy caterpillar: *Euproctis fraterna, Thiacidas postica* : Lymantriidae

The caterpillars in the initial instars remain gregarious and scrape leaves and tender fruits. Later instar caterpillars disperse and devour leaves, fruits and tender shoots.

Leaf roller : *Psorostichia (=Tonica) zizyphi:* Oecophoridae

The Larva is pale yellowish green with black head. They roll up the leaves, causing crinkling and defoliation. The adult is a small, brown moth.

Mealybugs and scales: *Drepanococcus chiton, Nipaecoccus viridis:* Coccidae;*Planococcus lilacinus:*Pseudococcidae;*Kerria lacca*: Lacciferidae

The waxy scale, *D. chiton*, causes extensive damage to ber in south India. They are found in large colonies, sucking sap from young shoots and tender leaves. Removal of sap leads to drying of shoots. *Kerria lacca,* the lac insect can damage in a similar way except that the colour of shoots becomes dark brown due to encrustation by scales.

Thrips: *Florithrips traegardhi* : Thripidae

These thrips, has been observed causing serious damage on ber leaves. Excessive feeding injury results in curling and drying of the leaves.

Chapter - 7

CITRUS

Citrus butterfly: ***Papilio demoleus, P. polytes*** **: Papilionidae**

The pest is distributed all over the world and has a host range that includes citrus and other rutaceae plants like *Aegle marmelos.* Adult has the head and thorax black with a creamy-yellow coloration on the underside of the abdomen. Wings are dull black, orna-mented with yellow markings and a distinct red patch on the hind wing. The underside is orange and yellow coloured and can be seen when it flirts for honey collection or just before egg laying. The Antenna is black and clubbed.

Damage

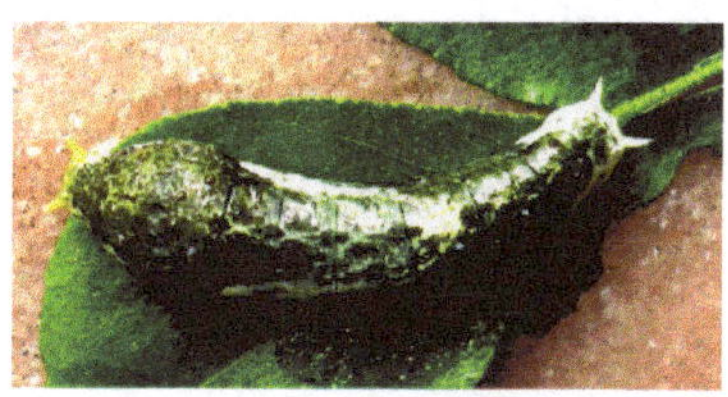

The young larvae are found on the upper surface of leaves and feed on the leaf lamina from margin into midrib. Grown up larvae feed on mature leaves and cause severe defoliation to the entire plant.

Biology

Egg is laid singly or in groups of 2-3 eggs on leaves. Up to 183 eggs/ female are laid and these hatch out in 8-6 days. Eggs are glued firmly onto the

surface of leaves and are pale or greenish yellow, when freshly laid, but later turn brown. Larva eats the egg-shell forming its first food. The larvae also eat their own exuviae after each moult. Larval period is 13-26 days in summer and four weeks during November to December. Young larvae resemble bird droppings and thus escape predation. Older larvae are yellow-ish green with a horn like structure on the dorsal side of the last body segments. When disturbed, the caterpillar pushes out a bifid purple structure called as **osmeterium** from the top of its prothorax that releases a peculiar smell. Full grown larva spins a girdle around its body and pupates on a twig as chrysalis which lasts for 8-11 days. Mechanical collection and destruction can be practiced against this pest, especially in small trees. Big trees do not get damaged to a great extent.

Stem borers: *Chelidonium cinctum; Chloridolum alcmene:* Cerambycidae

These are serious highly destructive pests in South India. The adults of *Chloridolum alcmene* are shining blue longicorn beetles about 25 mm in length. Adults of *Chelidonium cinctum* are dull metallic green to dark violet with a yellow band across the middle of the elytra. *Chelidonium* is a borer that pushes out powdered wood through bore holes. The stem and branches dry up.

The eggs are laid on twigs and thorns and covered by a resinous fluid secretion. About 30-50 eggs are laid by the female. Eggs hatch in 11-72 days. The grub which is creamy white with flat head bore for about 10 months. They bore into stem and feeds on the internal tissues, which result in drying of terminal shoots in early stages, followed by wilting of thicker branches and main stem. Symptoms of attack are seen as exudation of gum and accumulation of powdered wood on the ground below. The full grown grubs pupate in a tunnel for three weeks. They emerge in April-May. The total life cycle occupies one year. For management of the pests, cutting and burning of affected branches, injecting fumigants like Aluminium phosphide or injecting insecticides with fumigant action (Dichlorvos) and painting the stems with insecticide slurry with a WP formulation of 2-5% carbaryl or equivalent is recommended.

Orange trunk borer: ***Anoplophora versteegi*** **:F. Lamiidae.**

Adults are small to medium sized beetles with whitish elytra and long antennae. The egg is laid singly in cracks on the bark up to 2 feet height from the ground. About 35-85 eggs are laid per female. The grub feeds on sapwood making horizontal galleries before entering into the centre of the trunk resulting in death of the tree. Larvae last for 267-290 days. Pupation is inside the larval gallery for 33-43 days. Adult feeds on the leaf tissue along the midrib, keeping margins intact.

Management is by pruning the infested branches and also painting the trunk with carbaryl paste 0.5% or chlorpyriphos 0.3-0.5%. A spray on the twigs and branches can also be given, if the pest is endemic.

Bark caterpillar: ***Indarbela quadrinotata;Indarbela tetraonis*****: Metarbelidae**

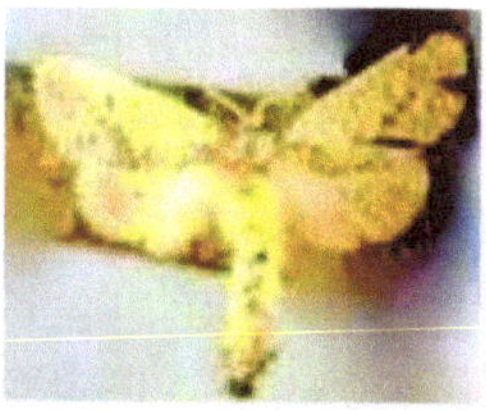

I.tetraonis is a pest with a host range that includes mango, guava, loquat, mulberry, pomegranate, drumstick, litchi, rose and a number of forest and ornamental trees. Adults are stout pale brown moths with grey wavy markings on the wings and a wing span of about 12 mm.

The eggs are laid in masses of 15-25 under loose bark of the trees or crevices in the bark during summer. There can be more than 200 eggs per female. Eggs emerge in 8-10 days and the newly hatched larvae are dirty brown and grown up caterpillars are pale brown with dark brown head. Larvae bore into the bark and sapwood making zig-zag galleries filled with silken webbed masses comprising chewed up materials and excreta. They feed during night and are sheltered in the webbed gallery of silk and frass.In severe infestation the flow of sap is interfered. The growth of plant is arrested and fruit formation is drastically reduced. The larvae when full grown, are big, measuring about 5-6 cm in length. They live for many months. Pupation is inside the larval tunnel for about 3-4 weeks. Adults emerge in summer and are short lived. There is only one generation / year.

I.quadrinotata also has similar habits. They also have a very similar life cycle, which is completed with one generation per year.

Management is by removing the silken galleries and spraying the bark with a strong contact insecticide like chlorpyriphos @ 5ml/l when the attack is first noticed. Spray should be done in the evening, when they come out to feed on the bark outside. Ensure that the webbings and excreta are scrubbed off and if possible, spray at dusk, on the emerging caterpillars themselves. Ensure systematic monitoring against the pest. A second spray may be required when the larvae are more in number and when the attack is severe and not managed by the first spray. Triazophos 40 EC @ 2.5 ml/l or fipronyl @ 2 ml /l are other options.

Fruit sucking /piercing moths: *Othreis materna, O. fullonica, O. ancilla, Anua coronata, Lagoptera honesta* : Noctuidae : Lepidoptera

Othreis materna

O. fullonica

O. ancilla

These attack citrus, mango, grapes and apple. The adult moths pierce the fruits for sucking the juice and make characteristic pin-hole damage in fruits. Bacterial and fungal infections take place at the site of attack. The whole fruit turns yellow due to the sap sucking, and it drops from tree and looks like a premature fruit. In severe cases, all fruits can be lost, but these are usually only minor pests. Adults of *O. fullonica* has a wing span of 9-10 cm and are grey with white patches and a demarcating line from anal tip to base making a triangular division of the wing. Hind wing has a black patch on the outer margin and a crescent moon black patch facing the outer edge in the middle of the yellow wing

O. materna have white and dark patches on the greenish brown forewings and a black spot on the hind wings.

O. ancilla has white bands in the middle of the grey fore wing with whitish outer margins. Hind wings are yellow with an outer black margin and a central black spot.

Anua coronata has a brown forewing with a bean shaped black spot. Hind wings are yellow with black patches on the outer and basal margins.

Lagoptera honesta also has brown forewings with abdomen and hind wings crimson coloured with a large black patch on the border

Eggs are laid on wild plants and weeds in and around citrus orchards. The egg period lasts for about two weeks. They are round and translucent. Eggs are usually laid on plants like Rangoon creeper *(Quisqualis indica),* Amrutha or guduchi and many weeds. Larvae are semi-loopers and have a stout appearance having a dorsal lump on the last segment of the body. They have distinct yellow / red eye spots on head/ lateral sides. Their velvety dark-brown background along with other patterns makes them cryptic. Full-grown larva, when disturbed, assumes a characteristic snake like posture by curving round head and raising hind part of its body. They have 5 instars completed in about 4 weeks. Pupa is in a pupal case by webbing together pieces of leaves and soil particles. Pupa is thick and dark reddish brown.

Management

- Destroy the weed hosts like Rangoon creeper
- Set up food lures with rotten tomatoes (or) pieces of citrus fruits baited with a contact poison like malathion 0.03%
- Cover the fruit with perforated bags wherever feasible
- Bait with fermented molasses / jaggery + malathion 1 ml/l

Citrus leaf miner: *Phyllocnistis citrella* : Gracillaridae : Lepidoptera

This leaf miner has a host range that includes citrus, pomello, willow, cinnamon and loranthus. Adult is a minute silvery white moth with 4.3 mm wing expanse with fringed wings and a black spot on the tip of the forewing. Larva mines into the tender leaves and form zig-zag galleries. They feed on the tissue in between the epidermal cells and make the area papery and crinkled. The leaves turn pale, papery, distorted and dry up. The mining of larvae may lead to infection by fungi and bacteria causing 'citrus canker'.

Biology

Eggs are laid singly inside the leaves and twigs and hatch out in 2-10 days.

About 70 eggs are laid per female. Larva is pale yellow / pale green with well developed mandibles. It is fully fed in 5-30 days. Pupation is in the leaves and can occupy 5-25 days.

Management

This is normally a minor pest and does not require an insecticide spray except when it is endemic and assumes epidemic status. At those times, spray Neem Oil 3 ml/l or dichlorvos 1 ml/l or carbaryl 4g/l or dimethoate 2 ml/l or fenthion 1 ml/l.

Leaf roller: *Psorostichia (=Tonica) zizyphi* : Oecophoridae

Adult is a small brown moth and the larva is a pale yellowish green caterpillar with black head. They cause rolling up of leaves, crinkling and defoliation.

Aphid: *Toxoptera citricidus, T. aurantii, Aphis gossypii, A.craccivora, Myzus persicae:* Aphididae

Toxoptera are generally specific to citrus and attacks at the time of flowering. The adults are shiny black aphids and their feeding causes yellowing and crinkling of tender shoots. Other species are polyphagous.

Citrus black fly: *Aleurocanthus woglumi* : Aleyrodidae

The black flies, under the family Aleyrodidae, have a host range which includes citrus, sweet orange, avocado, grape vine, mango, guava, pear and plum.

Adults are dark orange flies with smoky wings. Forewings have four whitish areas of irregular shape. Eggs are yellowish brown oval and laid spirally on the leaves. They hatch in 7-14 days. The nymphs are scale like, shiny black and spiny and has an outer white fringe of wax. They suck the plant sap from leaves and cause curling of leaves and premature fall of flower buds and developing fruits. They feed for 38-60 days and have a stage called as pupa which is oval and black. Their dorsum is arched with long black spines.

Management

Spray quinalphos 2 ml/l,Hostathion 40 EC 2.5-4ml/l, Imidachloprid20EC @0.5-1 ml/l, Thiacloprid 0.5-1 ml/l or Fipronyl 5%@2 ml/l.

Thrips: *Thrips nilgiriensis:* Thripidae, Thysanoptera

Adult is with yellowish, fringed wings. Feeding by the thrips causes ring like scarred area encircling the fruit stalk and irregular mottled patches on rind.

Shoot psyllid: *Diaphorina citri.* : Psyllidae

Adult is minute with long wings that extend beyond tip of abdomen. Wings and body is covered with a white waxy powder.

Both nymphs and adults suck cell sap from leaves, which curl over, dry and fall off.

About 200 oval pale yellow eggs are laid per female which are laid singly on the underside of soft young leaves. Nymphs are also pale yellow, with purple eyes and are marginally fringed with bristles. They live for about 25-71 days and form a pupa that is broadly oval, pale yellow with an orange yellow band in the middle of body.

Citrus red scale: *Aonidiella aurantii* : Diaspididae

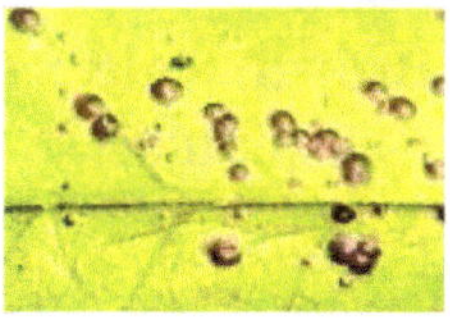

The hard scales congregate on the leaves and desap. These are not usually problematic. The pest is usually seen in gardens which are senile or uncared for. The first step for eradication would be to remove the affected branches and burn them.

Citrus fruit rust mites: *Eutetranychus orientalis, E.banksii, Panonychus citri:* Tetranychidae

Feeding by adults and nymphs causes silvery, scaly discoloration or rusty to black discoloration on the fruits.

Eriophyid mite: *Phyllocoptruta oleivora* : Eriophyidae

The adult eriophyid mites are yellow, and are flat. Feeding by adults and nymphs causes silvery, scaly discoloration on lemon fruits and rusty to black discoloration on other fruits; affected fruits are smaller and the skin of injured fruits are thicker.

Cottony cushion scale: *Icerya purchasi*:Margarodidae

This is the pest on which the classical biological control experiments were performed. In India this has not been a serious pest. This insect produces three generations per year and can overwinter in all stages. The scale is reddish brown about one cm long, and the females are covered with a white shell like ridged mass that can contain about 800 eggs inside. The nymph is red with black legs and has hairy growth and long antenna. They attach to the stems and bark and feed on the sap. For control, a dormant oil spray is recommended prior to the growth of the fruit. Bio control is the classical remedy.

Citrus fruit fly: *Ceratitis capitata*: Tephritidae

This is also not a serious pest in India. The affected fruits are rotten due to the feeding by the maggots.

Other pests that are seen in citrus include the following:

Whitefly : *Dialeurodes citri*: Aleyrodidae

Brown scale : *Chrysomphallus aonidium* :Diaspididae

Green scale : *Coccus viridis*: Coccidae

Mealy bugs : *Planococcus citri, P. pacificus* : Pseudococcidae

Shoot borer : *Oberea latteapicallis* : Cerambycidae

IPM in citrus

Monitoring and timely control measures are most important in Citrus IPM. Population of scales and mealy bugs, mites and thrips have to be monitored using sticky traps or yellow sticky traps or by visual monitoring. Differentiate between males of scales, parasitoids and other small insects and monitor for increase in pest population to manage the rising populations. Soft scales, black flies and whiteflies can be managed from reaching damaging populations by initial spot applications of Methyl parathion 50 EC 1.5 ml/l, or chlorfenvinphos 20 EC 2.5 ml/l or quinalphos 25EC 2.0ml/l. Against mealy bugs, provide a ring band of insecticide paste (methyl parathion).Crawlers are caught on sticky traps. Place them on shoot bearing the fruits at a length of 5cm. Release the predator, *Cryptolaemus montrouzieri*

a. Mark the mealy bug infested plants early in the season.

b. Put a band of quinalphos or methyl parathion dust @ 8Kg/ac around the tree trunk leaving about 30cm from the main stem to ward off ants that can carry the nymphs. Follow the ant control methods such as destruction of ant holes, red ant nests and skirting of citrus trees after fruit harvest which prevents the ant migration through side branches.

c. Release the Cryptolaemus beetles @ 10/tree once the patrolling of the ants on the trunk is stopped. In the initial weeks of the release, periodically check for the establishment.Put dried leaf mulch around the tree trunk 20 days after the release of the beetles to facilitate pupation of the full grown grubs.

d. Make one to three release per annum depending on the mealy bug populations:

Monitor for cottony cushion scale build up and the population of vedalia beetles. Shift vedalia to the sites of build up and give augmentative releases. In the case of leaf miner attack, or caterpillars, hand pick the larva and destroy. Fruit sucking moths develop on the weeds and hence destroy the weed host *Tinospora*. Use poison bait for adults (Dilute suspension of the fermented molasses + Malathion 0.05%. Set up light traps or food lure (pieces of poisoned fruits) to attract adult moths. Leaf miners may be managed, with chemical sprays like Dichlorvos 76 WSC 1ml/l, fenthion 80 EC 1ml/l,monocrotophos 36 WSC 1.5 ml/1,NSKE 5%, Neem formulations 2ml/l or Neem cake extract 5%. If leaf caterpillar becomes serious, spray malathion 2ml/l. Chemicals that can be used in rotation are Methyl demeton 25EC 1ml/l for aphids, or endosulfan 35EC 1.5 ml/l, carbaryl 50 WP 4g/l against shoot borer/stem borers. Rust mite populations can be managed by Dicofol 18.5 EC 2.5 ml/l or wettable sulphur 50 WP 2g/l. Nematodes can be managed with application of Aldicarb 10 G, or carbofuran 3 G or other granules.

Chapter - 8

CUSTARD APPLE

Mealy bugs: ***Ferrisia virgata, Planococcous lilacinus, P. pacificus, Maconellicoccus hirsutus, Perisopneumon ferox:*** **Pseudococcidae**

The fruit is seen covered with a mealy coating due to colonies of small reddish-white mealy bugs. They also infest leaves, shoots and buds. Due to sucking of sap by nymphs and adult female mealy bugs, fruits remain small in size and become shrivelled. Large number of mealy bugs is seen in the tip of the petiole abd also on the grooves of fruits epicarp. The fruit also becomes blackish due to the development of sooty mould. This affects its market value. *Maconellicoccus hirsutus* has pinkish nymphs and adults. Management has to start when the first colony of the mealy bugs are seen on the fruits that are set first, in the beginning of the season itself. A spray with a contact insecticide like quinalphos(25 EC) 2 ml/l, Hostathion 2.5-4ml/l, Imidachloprid(20EC) @0.5-1 ml/l, Thiacloprid(22EC) 0.5-1 ml/l or Fipronyl (5EC) @ 2 ml/l. Add liquid soap to the spray fluid @4-5 ml/l to dissolve the mealy coating, enable penetration of the spray, cause asphyxiation and better control of the mealy bugs. Insecticides are the last resort. See IPM for the crop at the end of this chapter.

Fruit- borer: *Anonaepestis(Heterographis) bengalella*: Phycitidae

The fruit borer, *A. bengalella* is another sporadic pest of custard- apple in south India. The adults are small moths. The caterpillars bore into fruits making irregular tunnels, covered with black excreta and damaging the mesocarp. Growth of infested fruits is affected and fruits drop. The adult is yellowish with a pale band on third tergite. When severe, the fruits may need a pesticide umbrella with very low residual insecticides like dichlorvos. . Insecticides are the last resort. See IPM for the crop at the end of this chapter.

Fruit fly: *Bactrocera (Dacus)zonata:*Tephritidae

The attack causes fruit rotting and fruit drop.

IPM for custard apple: This crop is generally managed without pesticides. Lookout for mealy bug infestation in the beginning of the flowering season itself. Application of fish oil rosin soap @25g/l or teepol @25 ml/ l will take care of the crawlers. This has to be done on the main stem and bark of branches, before the flowering season if the attack is endemic. For fruit fly attack, apply a bait spray of Malathion 2ml in a solution of 50 g jaggery plus 25 g fruit pulp per l from half ripe stage of the fruit. The sprays can be repeated at need based intervals from this stage onwards. Monitor fruit borer attack and remove and destroy affected fruits immediately.

Chapter - 9

FIG

Stem-borer: *Batocera rufomaculata, B. rubus, Olenecamptus bilobus:*Cerambycidae

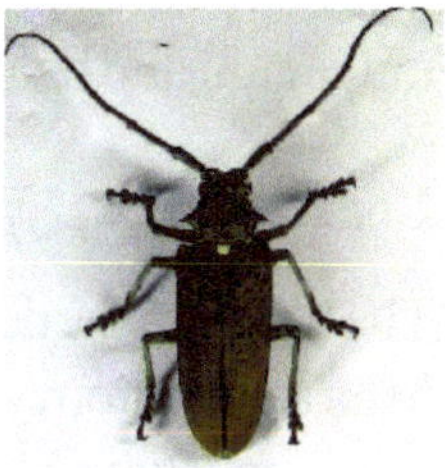

These are large dark brown adult beetles that lay eggs in loose bark. Grubs bore into the trunk and make tunnels in branches and main trunks. This causes drying and wilting of the branches. Adults of *O. bilobus* are pale whitish with long serrate antenna.

When bark is suspected to have the grubs, try mechanical removal of them. Paste the tree trunk with mud paste containing insecticides like chlorpyriphos or carbaryl. In endemic areas, spray the bark with Carbaryl 4gm /l, Malathion one ml/l, or Chlorpyriphos 2.5ml/l

Scale insects: *Parlatoria oleae* : Coccidae

The nymphs and adult females of the scale insects suck sap from flushes and fruits. Affected leaves and shoots show yellowing and drying.

Leaf roller: *Phycodes minor, P. radiata*: Glyphiterygidae

Greenish caterpillars of the leaf rollers feed on the epidermis of leaves in spring, and later on fold the leaves and feed within infested leaves. Larvae of *P.radiata* are yellowish white with dark lateral stripes. The adult is small greyish brown. Larvae of *P.minor* are light green with yellow, shiny head and lateral stripes.

Gall makers (midges): *Udumbarie nainiensis, Anjeerodyplosis peshawarensis* : Cecidomyidae

These are pests in north India and damages the flower buds by ovipositing in them. The maggot feeding causes discolouration, wrinkling and small protuberances on the fruit. *A. peshawarensis* oviposit in tender fruits. On hatching, the maggots feed on fruit pulp. The fruits become elongated, shrivel and drop prematurely.

Leaf caterpillars *Glyphodes* spp. Pyraustidae

The larva is greenish and cause defoliation. The adults are green or black or white coloured moths.

Wild silk worm: *Ocinara varians* : Bombycidae

The larvae are pale coloured and humped anteriorly with a slight horn at the tail region and hence resemble twigs. The adult is reddish grey.

Leafworms : *Hypsa ficus* : Hypsidae

The larvae cause defoliation. They are black with hairs arising from yellowish brown warts. Adults are yellow with black dots on the wings.

Thrips: *Giganotothrips elegans* :Thripidae

Nymphs and adults of these large thrips desap from tender leaves causing curling and drying.

Chapter - 10

GRAPE

Grapevine stem girdler: ***Sthenias grisator*** **: Cerambycidae**

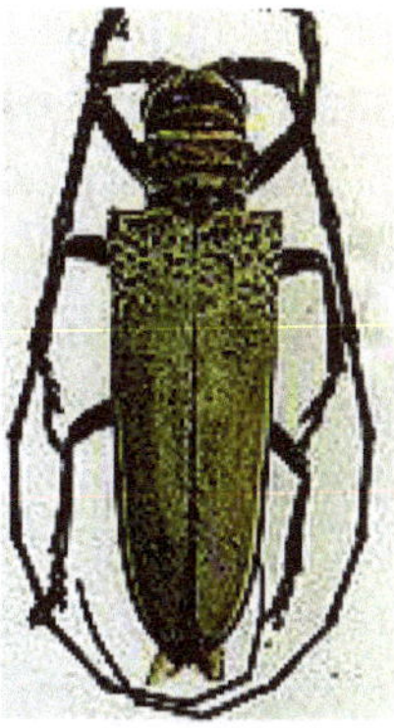

This is a pest that can also survive on mango, rose, cocoa, coffee, almond, and mulberry. The ovipositing adult causes wilting of branches and then the entire vine. Adults are medium-sized, grey coloured beetles with a white spot at the center of each elytron. They are active at night.

Eggs are deposited in clusters of 2-4, underneath loose bark of girdled branches. Eggs are oval and enveloped in a white parchment like covering. Adults emerge out and start ovipositing individual eggs at a point in the vine in a circular fashion so that the vine dries up above this point. Egg period takes about 8 days. The grub has dark brown head and mouth has a pair of prominent mandibles, each with two teeth. The grub continues its life as a semi saprophyte and pupates in the same tunnel. A single generation can occupy one year.

Grapevine flea beetle: ***Scelodonta strigicollis*** **: Eumolpidae**

The grubs of this pest damage the roots of grapevine. The adults cause small holes on the leaves by nibbling on them. Adult is a shining flea beetle with a metallic bronze colour and with black patches on the elytra.

The adults lay the eggs in groups of 4-7 beneath the bark or in groups of 20-40 in the soil. Around 220-550 eggs are laid by each female. Egg period lasts for 4-8 days and the emerging grubs take 34-45 days to reach pupal stage which is in the soil. Pupal period ranges from 7-11 days. One life cycle takes about 40-50 days and there can be 3-4 overlapping generations/year.

Grapevine thrips : *Rhipiphorothrips cruentatus* : Thripidae

This pest is also a pest of cucurbits, banana, ginger, cinnamon, rose and jamun. Their attack causes silvery white patches on leaves and the leaves also show black excreta. There is yellowing and withering in cases of severe infestation. Adult is minute, blackish brown, with yellowish wings. Female thrips can reproduce with or without fertilization with the fertilized eggs producing females and unfertilized eggs producing males. They lay eggs on the under surface of leaves by making small slits in plant tissues, placing one egg in one slit. A female can lay up to 50 eggs which are dirty white and bean-shaped. Egg period is 3-8 days and the young nymphs are reddish to yellowish-brown. Nymphal period lasts for 9-20 days. A pupal phase is recognised, seen on the leaves and can crawl away when disturbed.

Grapevine mealy bug : *Maconellicosus hirsutus* : Pseudococcidae

The adult females are pinkish with a sparse covering of mealy wax. Their attack causes crinkling and yellowing of leaves and rotting of fruits.

Eggs are orange in colour. The female egg cluster contains 350-500 eggs in a loose cottony ovisac, inside the female. Egg period can be up to 5-10 days. Crawlers are orange in colour, and the male nymphs have 3 and females have 4 nymphal instars.

Leaf Roller : *Sylepta lunalis* : Pyraustidae

Adults are brownish moths with wavy lines on the forewings. The larvae are pale green with short hairs on the body. They are leaf rollers, rolling the leaves and feeding from within.

Leaf caterpillar/Sphingid : *Hippotion celerio*: Sphingidae

Even though these are very occasional pests, they are big defoliators when they do occur. The adults are beautiful moths, red, brown and blackish, and with hooked antennae. The larvae are stout and green with an anal horn like process and also with eye spots on the thorax. Handpicking itself can manage the pest, but do it when the larvae are quite small itself.

Leaf miner : *Phylloccnistis toparcha* : Gracillariidae

Leaves become mined and crinkled. Adults are small greyish brown moths. They mine the leaves and cause crinkling of the leaves and vine tips.

Plume moth : *Oxyptilus regulus* : Pterophoridae

Larvae of these small plume moths with deeply cut plume like wings are small, pale green or pink with median red line and feed on the grape fruits.

Flat-headed borer : *Sphenoptera lafertei* : Buprestidae

Buprestids are jewel beetles and the small coppery black adult beetles (10-13mm) of this species feed on foliage, generally attacking stems and branches exposed to the sun. The plants become weak or die in case of severe infestation.

Fruit sucking / piercing moths : *Othreis materna, O. fullonica, O. ancilla, Anua coronata, Lagoptera honesta, Achaea janata* : Noctuidae

Othreis materna, ***O. fullonica*** ***O. ancilla***

The adults of these moths are occasional fruit feeders and cause rotting of fruits and premature fruit drop.

Defoliating beetles : *Adoretus lasiopygus, A. versutus, A. duvauceli, A. bengalensis* : Melolonthidae

The beetles feed on leaves from margins or as shot holes. In severe cases, the adults may defoliate the vines of grapes completely.

Yellow and red wasp

Wasps cause damage by puncturing ripe fruits and feeding on them. Honeybees are also attracted to punctured fruits.

IPM in grapes: Grapes, being consumed raw, allows only minimal use of insecticides. As a general rule, even when insecticide use is warranted, use botanicals or chemicals with low residual and mammalian toxicity. Monitoring and managing at initial pest build up is the rule. Biocontrol of thrips, aphids and mealy bugs should be promoted. Monitoring of these in the blooming stage and immediate remedies based on the surveillance reduces pesticide application. Release *Cryptolaemus montrouzieri* as a predator against mealy bugs and scales. In severe infestation, use Dichlorvos +fish oil rosin soap. Against stem girdler, deterrent chemicals / botanicals like neem oil/NSKE/garlic extract can be sprayed on the stem during the incidence season. Remove and destroy loose bark and plant parts after every pruning.

Chapter - 11

GUAVA

Spiralling whitefly : ***Aleurodicus dispersus*** **: Aleyrodidae**

A. dispersus originated from the tropical countries in Central and South America. More recently, it has been reported from Asia and Africa. In India, Andhra Pradesh, Karnataka, Kerala, Maharashtra and Tamil Nadu faces its problem. *A. dispersus* is a highly polyphagous species. Its wide host range includes many vegetable, ornamental and fruit crops, as well as numerous trees and shrubs, including *Capsicum*, *Citrus*, *Cocos nucifera* (coconut), *Euphorbia pulcherrima* (poinsettia), *Glycine max* (soybean), *Hibiscus*, *Lycopersicon esculentum* (tomato), *Mangifera indica* (mango), *Musa* (banana), *Persea americana* (avocado), *Prunus* spp., *Psidium guajava* (guava)and *Solanum melongena* (brinjal) etc.

Immature and adult stages of *A. dispersus* cause direct feeding damage by sucking plant sap, which can cause premature leaf fall. Indirect damage is due to the heavy production of honey-dew and white, waxy material produced by the insect. Sooty mould develops on honeydew and decreases photosynthetic activity. Plants are also disfigured and may be unmarketable. The nymphs and adults remain as a colony and also cause crinkling of the fruits. There is yield loss in quality and quantity In places where it occurs, *A. dispersus* is generally considered as a serious pest, causing crop losses. Virus transmission is apparently not known.

Dissemination : Natural dispersal can be ensured by flying adults. Over long distances, the pest has already showed its potential for spread, being introduced into many different parts of the world. Movements of infested plants or fruits can ensure long distance dissemination.

Biology : The adults lay hundreds of eggs within the colony on the undersurface of the leaves. The eggs are laid in a characteristic spiral shape and this gives the species its name as the spiraling whitefly. These emerge within a few days and the young ones emerge out. These are called as crawlers and are the moving stage of the pest. They move out, by actively crawling out and find out a suitable site for feeding. Once settled, they feed, and go for the first moult. They do not normally move out further in their life time. The next is the second instar nymph, and two more moults occur, the last one transforming the nymph into a so called puparia. This, in close up, is a very decorative piece, resembling a pure white decorated icing cake, with silver threads tastefully arranged and protruding out. All the nymphal stages are phloem feeders. The nymphs secrete a waxy, filamentous, curly material over their body .The adult fly looks like a miniature wax coated moth, and can mate and later fly off to other plants and get dispersed by air currents. They can lay eggs on fresh plants. The natural control of the pest is by predatory lady bird beetles like *Cryptolaemus montrouzieri* or lacewings or by parasitoids like *Encarsia* spp. Chemical control is difficult, but contact insecticides, like Quinalphos 2ml/l or malathion one ml/l , with addition of soap solution @ 5ml /l of spray solution or neonicotinoids like imidachloprid @3ml/10 l will give reasonable control. Reducing the temperature of the microclimate by spraying water on the undersurface of the crops, especially vegetable crops, reduces the pest.

Guava scale: *Chloropulvinaria(pulvinaria) psidii*: Coccidae

The soft scales infest the leaves and tender shoots of guava on the underside. The egg sac can be seen and the mother scale moves off leaving the egg sacs as a trail (See photo). Desapping in large numbers by the scales cause loss of

vigour for the plants. The larvae of the lycenid butterfly *Spalgis epius*, camouflaged as one of the nymphs can be seen eating the nymphs of the scale in their colony.

Other scales attacking guava include *Saissetia hemispherica* and *Lecanium viridis*.

Tea mosquito bug : *Helopeltis antonii* : Miridae

Adults and nymphs feed on developing fruits, petioles, tender shoots, and leaf veins causing necrotic lesions. On the foliage, brownish-black necrotic patches appear and on shoots, elongate streaks and patches. A gummy exudate may exude from the feeding punctures. The feeding also causes blisters and corky growth on fruits, widespread drying of shoots, inflorescence and flowers and shedding of fruits.

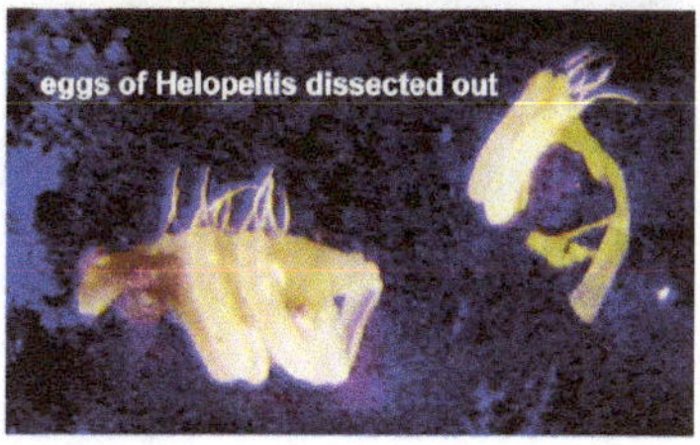

Around 30 eggs are inserted by the female into epidermis of the plant parts. They hatch in 7-8 days. Eggs are elongated, with a pair of hairs and an operculum. Nymphs live for 14-16 days and one life cycle is completed in 22-25 days.

Coreid bug : *Paradasynus rostratus* : Coreidae

This is not a main pest of guava, but is seen feeding on the tender developing fruits, which as a result dries up and drop prematurely.

Thrips : *Selenothrips rubrocinctus* : Thripidae

These are generally called as red banded thrips. They infest the leaves and causes browning and drying up of the leaves.

Bark caterpillar : *Indarbela tetraonis* : Metarbelidae (Arbelidae)

Caterpillars of these are bark borers that make galleries of silk and frass on the shoot surface and feed within.

Fruit borer: *Virachola(Deudorix) isocrates, Rapala varuna* : Lycaenidae

Both are lycaenid butterflies (blues and coppers). Virachola is bluish brown and Rapala is reddish brown. The larvae bore into the fruits.

Fruit borer : *Conogethes (Dichocrocis) punctiferalis* : Pyralidae

Adult is a medium sized moth with yellowish small black spots on the wings. The larva is pale greenish with pinkish tinge and fine hairs with dark head and prothoracic shield. They bore into the young fruits which dry up and fall. The bore holes are seen plugged with excreta.

Spray Dichlorvos or malathion 1.5 ml/lit of water if the attack is severe.

Fruit fly : *Bactrocera diversus, B. dorsalis* : Tephritidae

Bactrocera dorsalis

Maggots bore into fruits and feeds on soft pulp. The infested fruits show small sunken brown cavities when ripe. In immature fruits, there are dark greenish punctures and when cut open, the wriggling maggots are seen inside. The infestation causes rotting and dropping of fruits. Most of the development of the maggots occur after the fruit starts ripening and the pulp is rotten and inconsumable if the fruit is infested.

Egg is laid on soft skin of fruits. Larvae are pale cream maggots, cylindrical, 5-8 mm and with a conical head bearing the sucking apparatus. They cause the rotting of the fruit that drops to the ground and the maggots pupate in the soil for about 7 - 13 days. Adult is smoky brown with greenish black thorax having yellow marking.

Scarlet mite : *Brevipalpus phoenecis* F: Tenuipalpidae

Both nymphs and adults suck the cell sap from fruits which result in browning of the nodes and appearance of brown patches on the surface of fruits. In severe infestation, it leads to splitting of fruits. Total life cycle occupies about 22 days.

Other pests are

1. Aphids: *Aphis gossypii* Aphididae
2. Whitefly: *Aleurotuberculatus psidii* F: Aleyrodidae
3. Mealy bug: *Maconellicosus hirsutus; Ferrisia virgata* F: Pseudococcidae

IPM in guava

Monitoring the population of mealy bugs and whiteflies is to be done regularly. Plough the soil before every flowering period to expose pupae of fruit flies. Remove the fallen fruits and bury them deeply with lindane 1.3 D dusted over them. Release Cryptolaemus beetles @ 10/tree during the beginning of the season to prevent mealy bugs and whiteflies. Spray neem formulations@ 2ml/l at monthly/bimonthly interval at the time of flowering, if need be. Under heavy population, use Triazophos 2ml+ neemoil 5ml/l or phosalone 1.5 ml + neem oil 5ml/l. Against tea mosquito outbreaks, use malathion 50EC 2ml/l, endosulfan 35EC 2ml/l or fenthion 80EC 1ml/l.

Chapter - 12

JACKFRUIT

Shoot and fruit borer : *Diaphania (Margaronia) caesalis* : Pyraustidae

The shoot-and fruit-borer *D.caesalis* is a major pest of jackfruit in India. The larvae are reddish brown with pale brown head and thin hairs from black warts. The caterpillars bore into the shoots, buds and fruits. Affected shoots wilt or dry up and the fruits rot and drop. The adult is brown with grey elliptical patterns on the wings.

Bud weevil/leaf-eating weevil : *Ochyromera artocarpi, Onychocnemis careyae, Telurops ballardi* : Curculionidae

The small whitish grubs of these beetles bore into tender flower buds, terminal shoots and fruits. This induces premature fruit drop. The adults are greyish brown weevils.

Aphids : *Greenidia artocarpi, Toxoptera aurantii* : Aphididae

Both these species of aphids feed in colonies on tender leaves and shoots which get distorted and devitalized. The adult is a pale green aphid.

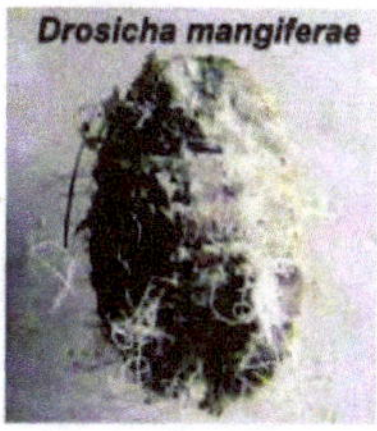

Giant mealybug : *Drosicha mangiferae* : Pseudococcidae

These are flat, oval, waxy white bugs observed in clusters on tender shoots and inflorescences, sucking the plant sap.

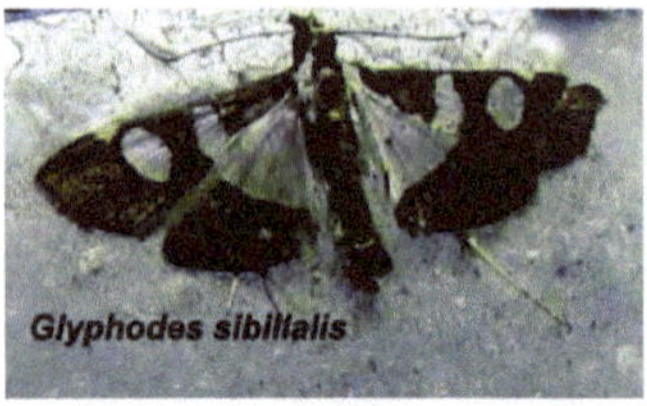

Leaf caterpillar : *Glyphodes sibillalis, G. bivitralis* : Pyraustidae

The larva is light yellow with black dots. Adult of *G.bivitralis* is light yellow with wavy lines. The larvae eat away the leaves.

Hairy caterpillar : *Perina nuda* : Lymantriidae

The adults are dimorphic moths; female with brownish white wings; male with half brownish and half transparent wings. Larvae cause defoliation

Lace wing bug : *Stephanitis chariesis* : Tingidae

Bugs of this species are with lace like wings. They suck the sap from undersurface of leaves.

Spittle bug : *Clavia lineaticollis, Cosmoscrata relata* : Cercopidae

These cause yellowing and terminal drying of shoots due to desapping by nymphs and adults.

Thrips : *Pseudodendrothrips dwivarna* : Thripidae

Whitish patches are seen on leaves due to desapping by nymph & adult. These are bicoloured thrips.

Chapter - 13

LITCHI

Fruit borers : *Criptophlebia illepida, Rapala varuna, Deudorix epijarbas, D. Isocrates* : Lycaenidae

Of the four species, *C.illepida* is more damaging to litchi fruits. The larvae bore into developing fruits and seeds, filling the bore-holes with excreta and the infested fruits start rotting.

Fruit-fly : *Bactrocera dorsalis* : Tephritidae

Maggots of *B.dorsalis* feed on fruit pulp and induce rotting. It has been observed as a serious pest in Bihar.

Eriophyid mite : *Aceria litchi* : Eriophyidae

The eriophyid is a major pest of litchi in India. Eriophyids are minute, worm like, microscopic mites with only two pairs of legs. It attacks new leaves, tender shoots and fruits. Feeding of sap initially cause yellowish-grey area which later on turn to chocolate brown. The leaves become twisted, curled and thickened. Heavy attack results in low fruit-set and poor fruit yield.

Bark-eating caterpillars : *Indarbela tetraonis, I. quadrinotata* : Metarbelidae

I.tetraonis is a pest with a host range that includes mango, guava, loquat, mulberry, pomegranate, drumstick, litchi, rose and a number of forest and ornamental trees. Adults are stout pale brown moths with grey wavy markings on the wings and a wing span of about 12 mm.

The eggs are laid in masses of 15-25 under loose bark of the trees or crevices in the bark during summer. There can be more than 200 eggs per female. Eggs emerge in 8-10 days and the newly hatched larvae are dirty brown and grown up caterpillars are pale brown with dark brown head. Larvae bore into the bark and sapwood making zig-zag galleries filled with silken webbed masses comprising chewed up materials and excreta. They feed during night and are sheltered in the webbed gallery of silk and frass. In severe infestation the flow of sap is interfered. The growth of plant is arrested and fruit formation is drastically reduced. The larvae when full grown are big, measuring about 5-6 cm in length. They live for many months. Pupation is inside the larval tunnel for about 3-4 weeks. Adults emerge in summer and are short lived. There is only one generation / year.

I. quadrinotata also has similar habits. They also have a very similar life cycle, which is completed with one generation per year

Management

By removing the silken galleries and spraying the bark with a strong contact insecticide like chlorpyriphos @ 5ml/l when the attack is first noticed. Spray should be done in the evening, when they come out to feed on the bark outside. Ensure that the webbings and excreta are scrubbed off and if possible, spray at dusk, on the emerging caterpillars themselves. Ensure systematic monitoring against the pest. A second spray may be required when the larvae are more in number and when the attack is severe and not managed by the first spray. Triazophos 40 EC @ 2.5 ml/l or fipronyl @ 2 ml /l are other options.

Chapter - 14

LOQUAT

Fruit –fly : *Bactrocera dorsalis* :Tephritidae

The process of egg laying and ovipunctures makes the fruits deformed. Feeding of maggots in the pulp induces rotting and fruit drop.

Bark-eating caterpillar : *Indarbela quadrinotata*

Severe damage on loquat due to *I. quadrinotata* was observed during October- November in Uttar Pradesh. There are some cultivars like Thames Pride which are less susceptible.

Loquat thrips : *Heliothrips* sp. : Thripidae

The thrips infest flowers during winter and suck the sap. Affected flowers get dried and do not bear fruits.

Scales : *Eulecanium tiliai, Parlatoria oleae* : Coccidae

The scales suck sap from lower surface of leaves and shoot which affects the vitality and fruiting capacity of the plants.

Chapter - 15

MANGO

Mango Fruit fly : *Bactrocera dorsalis* : Tephritidae

Bactrocera dorsalis

This is probably the most economically important pest of mango. It can attack many crops like mango, guava, peach, apricot, cherry, pear, ber, citrus, banana, papaya, avocado, passion fruit, coffee, melons, jack fruit and strawberry. There are also related species that may cross infect. The adults are light brown small flies with a single pair of membranous transparent wings. The other pair is modified into halteres. Eggs are thrust into the skin of the mangoes when they are semi ripe. A drop of resin comes out of the ovipunctures and can be seen as a black mark. While purchasing mango, this can be used as an indication of the presence of the fruit fly eggs. Eggs are laid in small clusters of 2-15 just beneath the skin of the ripening fruit. Upto 200 eggs are laid by a female fly during an oviposition period of one month. Eggs hatch in 2-3 days. The maggots are yellow in color and lots of them can be seen inside the mango pulp, causing rotting of the pulp. The fruit is rendered useless for human consumption. The infested fruits fall down and a liquefied pulp oozes down. The pupae

Fruitfly maggots in mango pulp

are seen in the soil. Adults emerge from the soil and starts ovipositing on new mango fruits.

Management

- Interspaces and the soil under the tree may be ploughed to expose and kill the soil borne puparia.
- Collect and destroy the infested and fallen fruits by deep burying, burning with dried leaves over it or put into refuse pits and malathion / other dusts dusted.

- Apply a bait-spray combining Malathion 50 EC @ 2 ml/l with molasses or jaggery (10 g/l) at weekly interval from mid ripening to harvest stage of mangoes.
- Coconut shell traps with crushed thulasi (holy basil, *Ocimum sanctum)* leaves with a drop of malathion put inside the sap is a good lure that attracts and kills both male and female flies in the immediate vicinity.
- Harvest the mangoes at 75-80 % maturity

Male annihilation technique

A pheromone trap commonly called as MAT trap to denote **Male Annihilation Technique** is kept in plastic bottles and hung in or in the periphery of the field. MAT blocks (Acacia blocks 3cm × 3cm/7mm impregnated with methyl eugenol dissolved in methanol) has been standardized and is commercially available in The Kerala Agricultural University. The trap is effective for about three months.

Hot brining

This is a new technique against the fruit fly, devised by this author. This is different from the hot water treatment hitherto recommended, and does not require a constant temperature facility to keep the mangoes in hot water for 30-40 minutes *etc.* Technique is as follows:

Harvest the mangoes at 75-80% maturity. (Not tree-ripe mangoes. The best flavor and sweetness will be for the mangoes at 75-80% maturity). This is the time when the mango fruits start developing a small dip just below the fruit stalk. This technique does not work when the mango is fully ripe. This is because, by that time, the fly eggs would have become maggots. You can't kill them easily.

Dip them in hot water prepared at the rate of ¾ th litre boiling water to one litre water. Add common salt at the rate of a small spoonful –around 2 g per litre of the solution. That would be a full scoop in a teaspoon for a 10 litre bucket. Let the mangoes be fully immersed in this salty water. If needed, keep a brick or wood piece to keep the mangoes dipped fully. Leave it to cool. After a few hours, take out, wash well in clean water, wipe dry and keep aside for a few days. When the mango is fully ripe, they can be eaten. The hot water and salt will kill the eggs just below the surface of the mango and also is anti fungal. So the mangoes will not rot due to disease or the fruit fly. There is also no dust or flowed ooze. So the mangoes are sweeter, without dirt, no fruit flies or rotting and uniformly ripe. They are cleaner too. Do not overheat. Do not add more salt. If some varieties are very soft skinned, use lower concentration of salt.

Mango hoppers : *Idioscopus niveosparsus, I. clypealis*

***Amritodes atkinsoni* : Cicadellidae I. niveosparsus**

These are commonly confined to mango alone. Out of the three, *I.niveosparsus* is usually the predominant one. The adult is wedge shaped with a tapering body and big head. They are light greenish brown with black and yellow markings and measure only 3-4 mm. *I. niveosparsus*: Three spots on scutellum and white band across the wing. *I. clypealis*: Two spots on scutellum and dark spots on the vertex. *A. atkinsoni*: Two spots on scutellum. Eggs are inserted individually into the young tissues on leaves, shoots, flower stalks and flowers. The nymphs pass through three instars and are full grown in about two weeks. The total life cycle is for about 3 weeks.

Both nymphs and adults suck the sap from tender shoots and inflorescence resulting in withering and shedding of flower buds and also wilting and drying of shoots and leaves. The flower stalks and leaves in infested trees become sticky due to the deposition of honey-dew secreted by the hoppers that encourages the growth of black sooty mould on foliage and other parts. During the off-season, the hoppers congregate on the leaves feeding on them. As a consequence, leaves become malformed and undersized. The hoppers take shelter in cracks and crevices on the bark during non-flowering season.

Management

If the attack is severe, spray two rounds of Acephate 75WP one g/L, (Waiting Period 15 days), Imidacloprid 20 EC 0.5-one ml/L, Thiacloprid 0.5-one ml/L, first at the time of panicle emergence and the second a fortnight later. This would usually suffice for the shoot pests including webbers and other flush feeders.

Stem borer: *Batocera rufomaculata*: Cerambycidae

This pest can attack mango, rubber, jack-fruit, fig, papaya, apple, eucalyptus and mulberry. Adults live for more than 6 months. These are greyish beetles with two pink dots and lateral spine on the thorax. Adults are 4cm long x 1.5 cm wide.

Eggs are laid singly on the bark or cracks and crevices on the tree trunk or branches. Eggs emerge in 1-2 weeks. Grub is yellow in colour and lasts for about 6 months. Pupation is in a pupal chamber excavated within the stem and is provided with an exit hole. Pupal period is 19-36 days.

The grubs feed by tunneling through the bark of branches and main stem. Initially many grubs are seen just below the bark where the eggs were laid. Later, they bore inside the main trunk. The damage may also show itself by the shedding of leaves and drying of terminal shoots in early stage of attack. When the damage occurs at main stem, whole tree succumbs.

Management

- Avoiding injuries at the base of trunk while pruning. Apply insecticide paste of Carbaryl 50 WP /Acephate 75 WP at the cut ends just after pruning.

- When bark is suspected to have the grubs, try mechanical removal of them. Paste the tree trunk with mud paste containing insecticides like chlorpyriphos or carbaryl.

In endemic areas, spray the bark with Carbaryl 4gm /l, Malathion one ml/l, or Chlorpyriphos 2.5ml/l

Mango nut weevil : *Sternochaetus mangiferae* : Curculionidae

Adults are stout built dark grey weevils, about 9x5 mm long. Eggs are pushed in singly on the marble sized fruits. The egg laying puncture gets sealed off as the fruit grows. Egg period is around 7 days. The grub hatches out inside the developing mango and when the stone is formed, gets inside, and starts feeding on the endosperm in the inside of the nut. The fruit is not affected when the mango is unripe. Larvae feed inside the stone and develop fully inside. The grub is creamy, yellow apodous with five larval instars. They also pupate inside the unripe mango. When the fruit ripens, they come out, and make a cut in the pulp, and goes out, leaving earth like excreta in the pulp. Only when the mango is used as a ripe fruit, the pulp is lost. Germination of the seed is also stopped and cannot be used for making seedlings for grafts.

If the pest is problematic, in endemic areas, a. spray with fenthion 100 EC @ 1 ml/l or carbaryl 4gm /l or malathion one ml/l can be given twice, first at the marble stage and the second a fortnight later.

Mango mealy bugs : *Drosicha mangiferae, Phenacoccus mangiferae, P.iceryoides* : Pseudococcidae

They have a wide host range, infesting crops like mango, apple, apricot, ber, cherry, *citrus* spp., fig, grape vine, guava, jack, jamun, litchi, mulberry and pomegranate. These are slow moving insects, and the nymphs and adults infest the leaves and inflorescence. Adults lay the eggs in soil and the crawlers move up and infest the tender leaves, shoots and fruits. They are sap feeders and usually are minor pests that don't need a chemical spray.

Scale insects: *Chionaspis vitis:*Diaspididae; *Ceroplastes rubens* :Coccidae

These also are desappers, and are usually minor pests.

Guava scale: *Chloropulvinaria(pulvinaria) psidii* : Coccidae

The soft scales infest the leaves and tender shoots of guava on the underside. The egg sac can be seen and the mother scale moves off leaving the egg sacs as a trail. Desapping in large numbers by the scales cause loss of vigour for the plants. In south India, this scale is preyed upon by the larvae of the lycaenid butterfly, *Spalgis epius*.

Bark caterpillar : *Indarbela tetraonis, I.Quadrinotata* : Metarbelidae

It damages the tree bark and feeds during night time.

Defoliators of the new flushes

Mango has many enemies when they flush and put forth new tender leaves. Many of them attack simultaneously, or one after the other. Most of them do a lot of damage to the young emerging very tender flushes.

Butterfly caterpillar: *Euthalia garuda*: Nymphalidae

The adult is a reasonably big butterfly about 7cm in wing span. The wings are brown with black and white markings on them. The caterpillar is of peculiar shape and has a series of branched leg like processes on both sides of the body and is colored green, so as to hide among the leaves. The larvae eat the leaves. No special sprays may be needed normally.

Flower caterpillar *: Rapala manea* : Lycaenidae

The adult is a dark brown butterfly with tails on hind wings and dark centered eyespots. Medium sized about 3 mm in wing span. Eggs are laid on the inflorescence and the young larvae feed on flowers and flower buds. The caterpillar is slug like and pupates on the inflorescence. Total life cycle until adult is 2+13+6=21 days.

Leaf miner : *Acrocercops syngramma:* Gracillaridae

he adult is a minute silvery moth. The larvae mine the leaf and cause white blisters on the new flushes.

Shoot webber : *Orthaga exvinacea*: Pyralidae

The adults are small grey moths around 3cm with dark patches on the wings. Eggs are laid on the petioles of very tender leaves or leaf veins. The larvae are gregarious initially and scrape the leaves. Later, they make tunnels of thin webs and take shelter within. The mature larvae are pale green with dark bands. A webbed cage of leaves can have many larvae inside. They pupate there and come out after around 11-14 days of pupal life. Spraying or dusting with insecticides kills the larvae.

Flower webber: *Eublemma versicolor* and *E. angulifera* : Noctuidae

The larvae are greenish yellow with light brown head. They also have a prothoracic shield. The female adult is purplish grey or pink. Spray phosalone 35 EC @ 2 ml/ lit.

Leaf/flush caterpillar: *Penicillaria(Bombotelia) jocosatrix*: Noctuidae

This is a medium sized moth with a wing span of 2.7 cm. It has brownish forewings with white lines and hind wings with a large white patch at the base. The caterpillar is green with stripes and spots in the body. Feeds on flushes and tender leaves.

Other noctuids include Shoot and flower caterpillar *Chlumetia transversa*

Hairy caterpillar: *Perina nuda* : Lymantriidae

The males of this moth have half brownish and half transparent forewings and the female is dull brownish white. They are defoliators.

Looper caterpillar : *Chloroclystis* sp. : Geometridae

This is also a moth with greyish wings with wavy lines. The brown loopers of this butterfly enter the bud and damage them.

Looper caterpillar : *Thalassodes quadraria*: Geometridae

The adults are small greenish moths and the larvae are brownish loopers that take on the pose of a leaf petiole for escape from predators.

Other geometrids are *Comostola laesaria, Gymnoscelis transversa, and G.imparatalis*

Thallassodes quadraria

Flush weevils

Leaf cutting weevil : *Deporas marginatus* : Curculionidae

These are small greyish brown to black weevils with a long snout and about 5 mm long. They lay their eggs on the undersurface of the young leaves on cavities scooped out near the midrib and then cut down the leaf at its base. Often, all the young leaves are cut this way. The grub on the fallen leaves mine them, complete their life cycle there and emerge out.

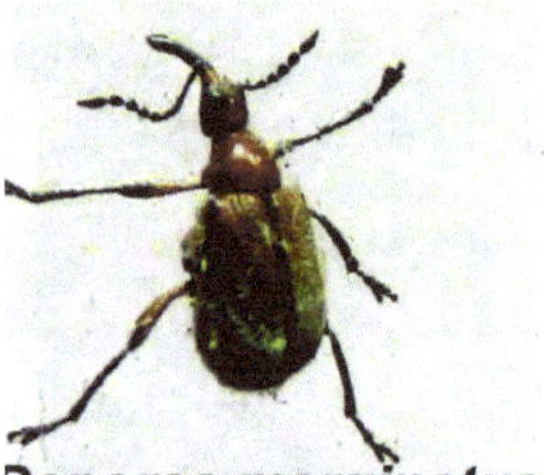

Deporas marginatus

Leaf twisting weevil : *Apoderes tranquebaricus* : Curculionidae

They are seen within the twisted leaf roll and feeding on it.

Leaf miner : *Rhynchoenus mangiferae* : Curculionidae

The adults are minute jumping weevils with thick hind femora.

They are leaf miners in the grub stage and many adults can be seen on the new flushes.

Leaf psyllid : *Microceropsylla (Pauropsylla) brevicornis* : Psyllidae

Adults of these greenish yellow psyllids with brown head and thorax lay the eggs on the tender leaves. The nymphs and adults cluster on young stems, leaf stalks and veins. They dry up. They also secrete honey dew and may allow for the development of sooty mould.

Flower gall midges : *Procystiphora mangiferae, Erosomyia indica, Dasyneura amaramanjarae*: Cecidomyiidae

The eggs are laid on the flowers by the adult midges. The adults and maggots are orange/yellow coloured. The maggots feed on the flowers and causes flower dropping and stunting and malformation of inflorescence. A spray of contact insecticide after emergence of the flowers will manage most of the flower feeders including midges, caterpillars, webbers, weevils and mites.

Spray dimethoate 30 EC or methyl demeton 25 EC @ 2 ml/lit.

Eriophyid mite: *Aceria mangiferae*: Eriophyidae

This is a very occasional pest on mango

IPM for Flush Feeders

All the flush feeders can be easily managed by an application of a commercially available contact insecticide. Though spraying is effective, it is advisable to go for dusts, because the dusts remain on the new flushes for a few days when new very tender leaves continue to emerge. The dusts and the repellency it creates will save the newly emerging leaves also, until they are hard enough to prevent further attack by the flush feeders.

IPM in Mango

Fruit flies are the major pests in Mango. Leaf and flower hoppers also cause yield reduction in Mango.

Set up Methyl Eugenol traps to monitor and kill fruit fly adults. The traps must be cleaned every week. The converging adults can be killed by adding a little from a solution of 2ml Malathion and 20 gm sugar dissolved in one litre water. This solution should also be sprayed in spots on the underside of the leaves of Mango and other vegetation to attract and kill adults. Collect all fallen fruits, and deep bury them after sprinkling lindane dust. Bait spray should be given at

10 day intervals from 50% maturity of fruits until harvest. The shell traps and pheromone traps arre also effective.

Mango hoppers, and other leaf and flower feeders like webbers, shoot weevils and mealy bugs are best managed by spraying Malathion@ 2ml per litre, Dimethoate 1ml per litre or phosalone 1.5 ml per litre. For flush worms and flush weevils, dust with any insecticide dust, just one to two days from the emergence of new flushes.

Chapter - 16

OLIVE

Olive-scale: *Metacernema japonica*: Black scale: *Saissetia oleae*: Coccidae

Olive scales are tiny pale-greyish insects which settle on branches, leaves and fruits and are active during April-November. *S. oleae* are round greenish brown scales. Nymphs and adults of both the species suck the sap from the tender shoots, and leaf petioles. Sooty-mould appears on leaves and branches of infested plants as a consequence of honeydew secre-tion. Shoots get dried up and fruits remain undersized.

Olive psylla: *Euphyllura longinova*: Psyllidae

These psyllids have waxy strands covering the body of nymphs, which give them a woolly appearance. They suck the sap from tender growing parts of the plants and fruit formation is affected.

Black tingid (lacewing) bug: *Eteoneus sigillatus*: Tingidae

The bugs feed from the ventral side of leaves. Due to continuous sucking of sap, yellowing of the leaves and sooty mould on the upper surface is seen.

Chapter - 17

PAPAYA

Fruit flies : *Bactrocera cucurbitae, Toxotrypana curvicauda* : Tephritidae

The fruit flies lay eggs on maturing fruits. The emerging maggots feed in the fruit pulp, causing rotting of the fruits.

AK grasshopper: *Poekilocerus pictus*: Acrididae

Nymphs and adults of the AK grass hopper feed on papaya leaves and tender bark. Maximum damage is done during July-August in North India.

Aphids : *Aphis gossypii, Myzus persicae*: Aphididae

Nymphs and adult aphids suck sap from leaves. They are more important as a vector of papaya mosaic virus disease. The affected trees die in an year or two.

Red spider mite : *Tetranychus cinnabarinus;* Scarlet mite: *Brevipalpus phoenicis*: Tetranychidae

The red spider mite is an occasional serious pest. Nymphs and adults suck sap from leaves which turn yellow and ultimately dry. In case of severe infestation, fruits are also attacked.

Stem-borer/fungus moth: *Dasyses rugosellus:* Tineiidae

Noted as a minor pest in Bihar, Maharashtra and Tamil Nadu, its caterpillars bore into the stems and feed internally.

Grey weevil : *Myllocerus viridanus* : Curculionidae

The adults of this grey weevil feed and skeletonise leaves in south India. The grubs are root feeders of papaya and other nearby plants.

The papaya Mealy bug : *Paracoccus marginatus* : Pseudococcidae

The papaya Mealy bug has created a serious problem for many crops and non crops in some southern states. The pest is called as a mealy bug because of the powdery coating or the mealy coating that is present over the body of the bug. Some farmers and common people have even described it as a fungal infection because of the white powdery mass over the insect and also because the insect does not move at all, once it develops into the adult stage.

The papaya mealy bug is a native of Mexico and/or Central America. In India, the pest was first reported from Coimbatore during 2008. Probably it was from here that the pest invaded the neighboring states. It has started a slow spread to many places, due to the tremendous amount of movement of men and material across the country. The damage is caused due to the sucking of sap from the veins of leaves of the affected trees, along with other parts like the fruits and flowers. The leaves become crinkled, lose their vigour, get yellowish, wither and then fall off. The adult female is yellow and is covered with a white waxy coating. The whole surface on the fruits, leaves or shoots where the pest is attacking, shows an fluffy, whitish, webbed white mass consisting of the egg mass, nymphs, pupae and adults of the pest. The whole plant gives an ugly look and withers eventually.

The pest has to be managed using insecticides, when the population is high.

But the doses required are high. This is because of the waxy coating of the mealy bug. Hence, it adds to the pollution of the environment and requires repeated application. Experiments have proven that application of insecticides like Malathion (2ml/l) or imidachloprid (3ml in 10L), in mixture with liquid soap, (A shampoo sachet (8ml) suffices for two litre spray fluid) helps to dissolve the wax, and ensures better pest control. Since the eggs in the ovisac continue to emerge for another 10 days, it is necessary to give a repeat spray after 10 days. Addition of liquid soap is a must for better efficiency of spraying, since otherwise, the waxy coating prevents the insecticide from penetration on the pest. Addition of powdered neem cake in the soil, would help management further.

In 2010, the parasitoids *Acerophagus papayae* (Noyes and Schauff), *Anagyrus loecki* (Noyes and Menezes) and *Pseudleptomastix mexicana* (Noyes and Schauff) were imported from Puerto Rico. All three species were screened in quarantine facilities and environmental assessments were completed. The National Bureau of Agriculturally Important Insects (NBAII), and various agricultural universities has been identified to function as the nodal agency for mass culturing and distribution of the parasitoids.

IPM in papaya: Generally papaya is not sprayed. A good monitoring mechanism and timely mechanical control will eradicate most pest problems. Aphid build up has to be prevented by systematic monitoring and timely control by mechanical removal of colonies and application of botanicals. When mosaic appears, remove and destroy the affected plants immediately. Against *Paracoccus,* use the parasitoids *Acerophagus papayae* or any other available effective parasitoid.

Chapter - 18

PEACH AND PERSIMMON

Green peach aphid/peach leaf curl aphid : *Brachycaudus helichrysi* Aphididae

These are green aphids that suck the sap from growing buds. Leaf buds become weak and result in poor setting and premature fruit fall.

Blossom thrips : *Thrips rhopalantennalis, T. flavus, T. florum, Haplothrips tenuipennis, Frankliniella dampfi* : Thripidae

These are tiny, yellow, pale-brown and black slender insects seen within the flowers and flower buds. Adults and nymphs lacerate floral parts and suck sap resulting in brown spots. Infested flowers give sickly and faded look. Lesions cause flower fall, inferior fruits and loss in yield. Attack on tender leaves causes necrotic silvering, curling, deformation and withering.

Peach fruit fly : *Bactrocera zonatus* : Oriental fruit fly : *B. (Dacus) dorsalis* : Tephritidae

Bactrocera dorsalis

The adults are brown to dark brown flies with ferruginous thorax and measures 7mm in length and 14mm across the wings. The females have tapering abdomen with a pointed ovipositor. Adults lay eggs inside fruits and the emerging maggots feed on pulp. The inside pulp gets tunneled due to active maggot feeding and becomes soft, gets fermented and drops. The attack by the maggots reduces yield and quality of the fruit.

Flat-headed borer: ***Sphenoptera lafertei:*** **Buprestidae**

Small coppery black adult beetles (10-13) feed on foliage. These generally attack stems and branches exposed to the sun. The plants become weak or die in severe infestation.

White Peach Scale: ***Pseudoulacaspis pentagona*** **:Diaspididae**

Nymphs suck the sap from leaves during April-May. The adults puncture ripe fruits during June-July. Due to the attack, the punctured spots become rotten.

Peach leaf roller: ***Cacoecia epicryta*** **:Tortricidae**

The larvae of the leaf roller web the leaves together. The adults are bell shaped moths while at rest as is typical of tortricidae.

Bark eating caterpillar: ***Indarbela quadrinotata, I. tetraonis:*** **Metarbelidae**

The gregarious caterpillars feed upon the bark surface as well as on stem, making galleries of silk and frass on the bark surface to serve as shelter

Termites: ***Microtermes obesi, Odontotermes (Termes)obesus*** **:Termitidae**

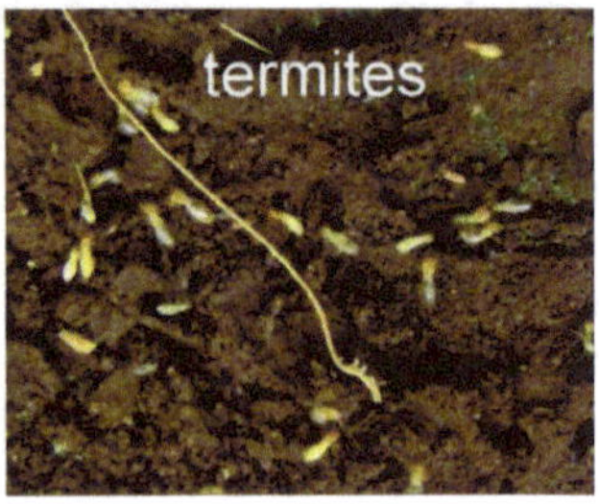

This is more a pest in dry weather. The termites feed on roots and stem bases near the ground level of almost all types of fruit trees in warm localities with sandy loam soil. The severely infested trees often dry and yield is reduced

PERSIMMON

Leaf roller: ***Hypocala rostrata:*** **Noctuidae**

The dark grey and orange moths of this species lay eggs on leaves in early-July. The Caterpillars fold leaves, feed inside and pupate within the leaf folds. These insects remain active throughout the rainy season.

Chapter - 19

PINEAPPLE

Mealy bug: *Dysmicoccus brevipus:* Pseudococcidae

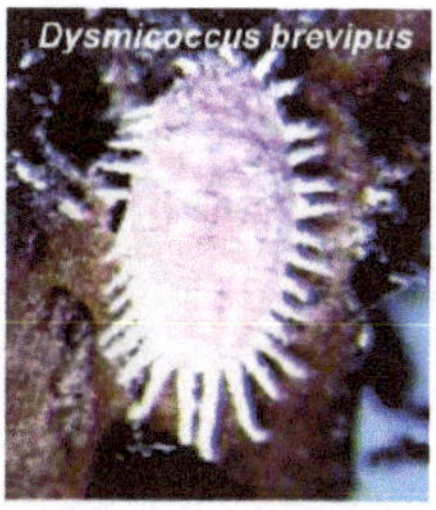

Mealy bugs are major pests of pineapple in India. They can attack roots and aerial parts. Colonies of reddish – white mealy bugs infest roots, crown, and fruits and suck sap. The leaves change in colour from green to red or pink and the margins reflect inwards and leaf tip starts drying. Severely infested plants become stunted and bear only small fruits. The attacked plants should be drenched with an insecticide solution of contact insecticides like Quinalphos or Dichlorvos or Chlorpyriphos@ 0.05% concentration, in between the fronds and allowed to drip down to reach the roots. Addition of liquid soap @4 ml per litre or 4 g of soap powder per litre is a must to get the best results.

Fruit-eating beetle/Corn sap beetle: *Carpophilus dimidiatus*: Nitidulidae

The beetles feed inside ripe fruits. Damaged fruits rot due to infection of pathogens.

Termites: *Odontotermes* spp. : Termitidae

Termite infestation is noticed in plantations in sandy soil. This is more a pest in dry weather. The termites feed on roots and stem bases. Drenching with insecticides for mealy bug control also takes care of termites.

Chapter - 20

PLUM

Plum weevil : *Amblyrrhinus poricollis, Myllocerus laetivirens, M. sabulosus, M. undecimpustulatus* : Curculionidae

Adult beetles feed on the epidermis of fresh and tender leaves leaving behind only the network of veins and affected leaves dry up and fall. They cause serious damage in young plantations.

Defoliating beetles

Adoretus duvauceli, A. versutus, Anomala dimidiata, A. flavipes, A. lineatopennis, A. rufiventris, Holotrichia longipennis, Hilyotrogus holosericeous, Lachnosterna coriacea, L. crinicollis, L. longipennis, Mimela passerine, M. pectoralis.

Green melolonthidae: *Anomala dimidiata*: Melolonthidae

These are pests in adult stage, where the different coloured beetles (Reddish brown, blue-black, and green), feed on foliage and developing fruits. The adults appear in May- June and lay eggs in soil at dusk during rainy season. They attack all temperate fruit plants whose leaves get skeletonized. The small fruits are eaten and fall prematurely. Their immature stages (white grub) feed on almost any root; crops, weeds, potato, vegetables and grasses. Normally the attack can be ignored, but when very severe, chemical control may be necessary.

Scale insects: *Eulecanium tiliac; Eriochiton theae*: Coccidae

The scale insects desap the plant parts and cause yellowing and wilting

San Jose Scale : *Diaspidiotus(=Quadraspidiotus) perniciosus*: *Hemiberlesia rapex, H. lataniae* : Diaspididae

San Jose Scale: *Diaspidiotus(=Quadraspidiotus) perniciosus*: Diaspididae

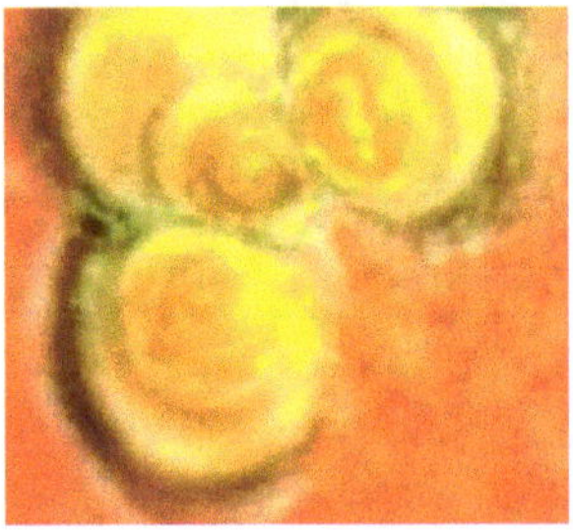

The San Jose scale derives its popular name from San Jose, California where Comstock discovered and named it in 1881. It has been considered the most pernicious (injurious, destructive) scale insect in the United States. The adult female is round, slightly convex with a black pustule. The adult males are small and winged. The body of adult female is yellow and is covered with a rounded dark gray scale up to two mm in diameter. Eggs are not seen because the females give birth to young ones. Yellow crawlers are born viviparously and emerge from the inside at the rate of two or three a day. The crawlers then move out to disperse to other parts of the plant, find out new feeding sites and start feeding. They moult after about ten days and begin to lose their eyes, legs and antennae. The adult female appears after the next moult and the scale develops, incorporating the larval exuviae. The development of the male involves three moults. The male nymph is more elongate than the female and the adult male is orange coloured and has wings. It lives only for a few hours.

The nymphs and adults of this insect congregate in large numbers on the bark and desap the trees. Later infestation may spread to the new flushes, blossoms and fruits. These parts are also shrunken, blemished and damaged due to the desapping. The infested region in bark becomes reddish pink and when the fruits are attacked, they have purple colouration.

Over wintering nymphs become active in mid March and the males emerge in April. Females are ovo -viviparous and reproduce in mid May. About 200-400 nymphs are released by each female, in a month. Crawler period is about 20 days. Total life cycle occupies 35-40 days. 4-5 generations are seen in a year before they start hibernation in mid October.

Management involves use of oil sprays and biological control. Pheromone traps are used for monitoring of the pest. Use dormant oil during the dormant period. Insecticide Dimethoate 0.05%, chlorpyriphos 0.05% or carbaryl any newer molecules such as spinosad, active against sucking pests can be used. If insecticides are used, the timing is important. Monitor the pest for the emergence of crawlers, otherwise called as the first instar nymphs, and apply the insecticides in the morning or evening, preferably with the addition of liquid soap, as described against mealy bugs and scales.

Plum fruit moth: *Cydia funebrana:* Tortricidae

The larvae of these plum moths bore into developing fruit and feed on the pulp. The fruits will ultimately rot and fall down.

Hairy caterpillars: *Euproctis fraterna, E. signata* : Lymantridae

These occur sporadically; caterpillars are profusely hairy. They feed voraciously and defoliate trees.

Leopard moth:*Zeuzera multistrigata* : Cossidae

The adults are white moths with attractive patterns, seen at dusk during May-June. The eggs are laid singly in cracks of barks. Within about a month, the emerging pinkish- white young caterpillars bore into branches and stems. The early instars scrap off the cambium and induce gummosis and later instars enter into the hardwood and make tunnels in the terminal shoots, secondary branches and main stem. This can result in drying of shoots within 1-2 years and of the whole plant within 3-4 years.

Chapter - 21

POMEGRANATE

Fruit borer (Anar butterfly): ***Deudorix(Virachola) isocrates*; *D. epijarbas:* :Lycaenidae**

This is a pest distributed all over India and adjoining countries .It has a host range covering apple, gooseberry ber, citrus, guava, litchi, loquat, peach, mulberry, pear, sapota and tamarind. Adult is a beautiful glossy bluish to brownish violet butterfly with an orange patch on forewings and a tail in the hind wing.

Larvae bore inside the developing fruits and feed on pulp and seeds. The fruits show round bore holes on the rind. Many caterpillars are seen in a fruit. Infested fruits get rotten by bacteria and fungi, fall off and give off an offensive smell. Egg is shiny white oval and is laid singly on calyx of flowers and on tender fruits. Eggs hatch in 6-9 days. Larval period is for about 18-40 days. Caterpillars are dark brown with lighter colored patches and having short hairs all over the body. They come out of the fruit, secure the fruit stalk to the stem and go inside to pupate. Occasionally pupation can be on the outside of the fruit also, attaching themselves to stalk of fruits. Adults emerge in 7-25 days. Covering the developing fruits with paper or muslin bags gives good protection to the fruits. The extent of loss varies from 20-80% depending upon the area, variety and season. This is a beautiful blue butterfly. Normally chemical control is not resorted to.

Fruit borer: *Rapala varuna*: Lycaenidae

This is another species of fruit borer in the same family, whose adult has orange coloured wings with marginal markings. The damage and biology is similar to that of Virachola.

Mealy wing /Whitefly: *Siphonimus phillyreae*: Aleyrodidae

The nymphs of this whitefly are bluish black. The adults and nymphs are seen congregating on ventral side of the leaves and suck the sap from them. This causes curling, yellowing and drying. Nymphs cause more damage due to desapping.

Mealy bug: *Planococcus lilacinus, Planococcus citri ,Ferrisia virgata:* Pseudococcidae

The nymphs of the mealy bugs infest flower buds and fruits in large numbers and desap. Severe infestation leads to dropping of flower buds and low fruit-set. The fruit becomes unmarketable. If non-chemical methods fail, an application with quinalphos @ 2ml /l or imidacloprid at 3-5 ml/10 l can be sprayed. Harvest the fruits only after 20 days. Add liquid soap or teepol 4ml/ l in the insecticide solution for effective kill of the mealy bugs.

Thrips: *Retithrips syriacus, Rhipiphorothrips cruentatus:* Thripidae

These are seen on the leaves and the attack causes discoloration and crinkling of the leaves. Both nymphs and adults lacerate tender leaves in margins and suck the exuding sap .Due to the attack, silvery white patches with yellowing and withering are noticed on the leaves.

Provide sprinkler irrigation or water spray during hot months. If the attack is severe spray Imidachloprid 20 EC at 2.5-3ml/10 L or Thiacloprid 0.5-one ml/L or Fipronyl 5% at 2 ml/L or Endosulfan 35EC at 3ml/l.

Stem borer: *Zeuzera coffeae* :Cossidae

The caterpillars of this pest bore into the branches and stems of pome-granate. The stem is weakened and sometimes dries up.

Bark borer: *Indarbela tetraonis;* Bark-eating caterpillar: *Indarbela quadrinotata* :Metarbelidae

Their larvae feed on bark under webbed galleries of silk and excreta. The attacked plants show the presence of such galleries on the bark surface.

Bagworm: *Clania crameri; Pteroma* sp. : Psychidae

Female of bagworm ***Clania crameri***

See the head protruding out of the case

The larvae of this pest feed on the surface tissues of the leaves. The males are winged and females are worm like. They make a house out of cut pieces of bark and carry it when they move. It is a mobile house and the larvae just protrude the head outside and feeds. Usually these are minor pests.

Slug caterpillar: *Latoia (Parasa) lepida*) : Cochlididae

These feed on the leaves of pomegranate.

Fruit sucking moths: *Othreis ancilla,O. cajecta, O.fullonica, O.materna:* Noctuidae

The adults of these moths puncture mature fruits during dusk time by inserting proboscis and suck the juice. Through feeding punctures secondary pathogens enter and cause fruit rotting. Fruit infestation varies from 40 to 90%.

Castor caterpillar: *Achaea janata* : Noctuidae

Semilooper feeds on leaves while the adult moth pierces the fruits with its proboscis and feed on the fruit, causing injury on the surface of fruits. The semilooper of *Achaea* also feed on the seedlings occasionally.

Pomegranate Aphid: ***Aphis punicae*****: Aphididae**

This aphid is serious on new flushes. The aphids infest tender shoots, leaves, flower buds and fruits. Severally infested leaves become curled.

Hairy caterpillar: ***Euproctis fraterna*****;** ***Porthesia scintillans*****: Lymantridae**

The hairy caterpillars are occasional defoliators

Other pests that are occasional pests include

Fruit fly: ***Bactrocera zonata*** **: Tephritidae**

Fruit borer: ***Conogethes punctiferalis*** **: Pyraustidae**

Non-insect pests

Eriophyid mite: ***Aceria granati*** **: Eriophyidae**

Both nymphs and adults live inside the rolls at the edges of leaves. Infested leaves become linear and deformed.

Red spider mite: ***Tetranychus punicae*** **: Tetranychidae**

This mite also feeds on the leaves, causing crinkling and distortion.

Chapter - 22

STRAWBERRY AND WALNUT

STRAWBERRY

White grubs and cut worms

These insects cut the roots and stems of young plants.

Hairy caterpillars

Defoliate plants during June-July

WALNUT

Walnut weevil: *Conotrachelus juglandis* : Curculionidae

The adult weevil is dark brown with white markings and has a prominent snout. They excavate circular holes on the fruit surface due to feeding. Grubs bore into the fruits and reduce the kernels into a black mass, and attacked fruits drop prematurely with grubs inside.

SECTION–II: MEDICINAL AND AROMATIC PLANTS

Chapter - 23

AMBRETTE (*ABELMOSCHUS MOSCHATUS*)

Spotted boll worms/ Shoot and fruit borer: *Earias vitella (=E.fabia), E. insulana*: Noctuidae

This is a pest on cotton, bhendi, hibiscus, holly hock and other malvaceous vegetables. Moths lay eggs singly in leaf axils, bracts and leaf veins and on under surface. Larvae bore into tender shoots, buds, flowers and fruits and feed on internal contents. Damage results in drooping of shoots and shedding of buds and flowers. The infested fruits have bore holes, are deformed and become unfit for consumption. Bore holes are plugged with excreta.

Adults of *E. insulana are* medium sized, with head and thorax whitish; forewings are pale white with a broad wedge shaped horizontal green patch in the middle that flows out to the periphery of the wing. The hind wings are cream white in colour. In the case of *E. insulana,* the adults are smaller than *E. vitella*. Head and thorax are uniformly pea green in colour and forewings uniformly pale yellowish green. They lay about 400 eggs / female. Spherical, light bluish, green, crown shaped, eggs are laid singly on shoot tips, buds, flowers and fruits. The egg period is about 3-5 days. Larva grows for 10-17 days and has six instars. *E. vitella* is brown, with longitudinal white stripes on dorsal side as against cream coloured body with orange dots on prothorax for *E. insulana*. Pupa lasts for 6-10 days and they pupate in a dirty white boat shaped cocoon on the plant itself.

Earias vitella **adult and larva**

E.insulana

Management

- Collect and destroy infested shoots, buds, flowers and fruits.
- Avoid cultivation of alternate hosts like cotton and hibiscus
- Release the lacewing *Chrysoperla carnea*.
- Setup light traps to monitor the moths and their egg laying.
- Spray extracts of other aromatic plants to deter the pest from egg laying and feeding.

The cotton aphid :*Aphis gossypii*: Aphididae

Both nymphs and adults cause the damage by sucking the plant sap from shoot tips, and leaves. Severe infestation may result in complete drying of affected parts. Attacked leaves curl up, wither and fall. These are yellowish green or brown coloured aphids.

Management

- Use entomopathogenic fungus *Fusarium pallidoroseum* to cause epizootics in aphids in the cowpea field.
- Apply water jets to the spots of congregation of the aphids
- In severe attack, spray 0.05% malathion or dichlorvos with a spray fluid of 500 litre water per ha just on the spots of young shoots where the aphids congregate. This pest has a quick multiplication rate and hence after the insecticidal treatment, water jets have to be repeated as soon as aphid population is found to have built up again.

Red spider mite: *Tetranychus cinnabarinus (= T. urticae = T. telarius = T. cucurbitacearum)*: Tetranychidae

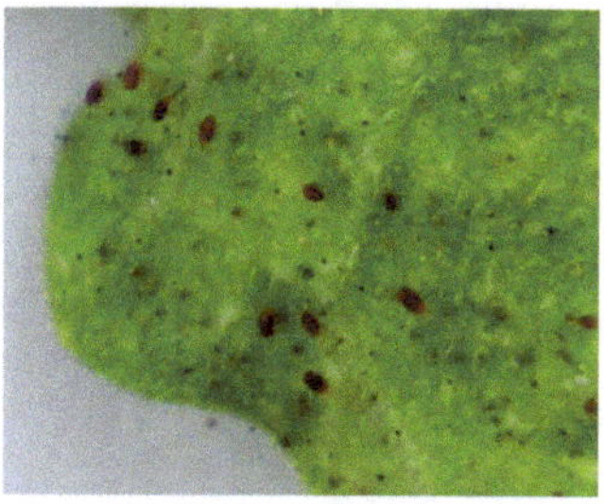

These are carmine red coloured small arthropods that move actively on the under surface of leaves. They spin webs and live in colonies beneath the webbing. Continuous feeding by the mites results in development of yellowish spots on the feeding sites and ultimately turns brownish. Mites penetrate the plant tissues by surface scratching by the mouth parts and suck the exuding sap. The leaves lose their turgor and the photosynthetic efficiency is lost and later dries up fully.

Management

During summer, apply water jets on the growing tips as a regular practice. Application of wettable sulphur 80wp 2g/l or Dicofol or Dimethoate 30 EC1.5ml/l or Phosalone 2ml/l on the under surface of the leaves may be necessary as a onetime measure when the pest assumes epidemic proportions. Provide copious irrigation on the under surface of the leaves.

Chapter - 24

ASWAGANDHA

Withania somnifera, also known as **Ashwagandha, Indian ginseng, Winter cherry, Ajagandha, Kan-aje Hindi, Amukkara in** Tamil and **Samm Al Ferakh,** is a plant in the ***Solanaceae*** or nightshade family.

Grasshoppers : *Trilophida an- nulata* : Acrididae

The nymphs and adults of this grasshopper feed on the margins of leaves of Aswagandha. It is only a minor pest, nibbling on the leaves.

Long horned grasshoppers : *Phenoropiara gracillis* : Tettigonidae

The adults and nymphs feed on the leaves in a shot hole manner. These are also easily controlled by hand picking and destruction.

Mealy bug: *Coccidohystrix insolitus*: Pseudococcidae

The nymphs and adults remain on the under surface of the leaves and suck the sap. Affected plants turn yellow, wilt and dry. Honey dew secreted by these mealybugs favour development of sooty mould on the areas where the honey dew falls.

Leaf hoppers: *Penthima* sp. : Cicadellidae

The nymphs and adults congregate on the leaves and suck the sap, causing yellow patches.These are only occasional pests, appearing in some years.

Cow bug : *Gargara mixta* : Membracidae

The nymphs and adults of this bug are seen feeding on the shoots, causing yellowing and devitalisation. The bugs are small and move towards the side when disturbed. They also have a dorsal horn like shield.

Sphingid larvae : *Dilephila nerii* : Sphingidae

Larvae feed voraciously on leaves resulting in severe defoliation. The larvae are common on the leaves and are big ones eating large quantity of leaves. The larvae are large and hence can easily be mechanically collected and destroyed by burning.

Epilachna beetle/Hadda beetle : *Henosepilachna viginti octopunctata, H. dodecastigma /H.Septima / Afidenta misera* : Coccinellidae

Adults of these species are distinguishable with the number of spots on the elytra. *E. vigintioctopunctata*: 14 spots on each elytra, deep red. Adult of *E. dodecastigma*: Copper-coloured, 6 spots

Both adult and grubs scrape the lower epidermis of leaves in characteristic manner leaving behind riddled areas. The leaves give a grazed appearance. In

Adult

Grub

severe infestation all the leaves may be eaten off leaving only the veins intact (Skeletonization), with the leaves or whole plants withering.

Henosepilachna viginti octopunctata

Adults lay 120-460 eggs/female. They are cigar shaped, yellow and are laid in clusters on lower leaf surface. The eggs hatch in 2-4 days .The grub period is 10-35 days. They are yellowish bearing six rows of longitudinal spines with terminal branching. Pupation is for 5-6 days and is seen on the stem or leaves. Total life period occupies 20-35 days with 7 -10 generations / year.

Management

- Systematically collect and destroy leaves with egg masses, grubs and pupae and also adult beetles
- Spray carbaryl 50% WP 4 g/l or endosulfan 35 EC 2 ml / l or malathion 50 EC 2 ml /l when populations explode.
- Diflubenzuron or Buprofezin also works well on the beetles
- However, chemicals to manage the pests of medicinal plants should be resorted to only as a last remedy.
- Taxonomists reveal that the number of spots is not a character to identify the species. It is clarified that the species feeding on brinjal is *Henosepilachna vigintioctopuncatata,* on cucurbits, it is *H.septima* and on pulses and cowpea, it is *Afidenta misera*

Ash weevil/ Mylloceros weevils : *Myllocerus viridanus, Myllocerus laetivirens, M. undecimpustulatus, M. discolor, Amylyrrhinus poricollis* : Curculionidae

M.viridanus is green and *M.discolor* is ash coloured with spots. The adults and grubs cause damage. Young larvae (grubs) feed on the roots of plants. The adult weevils cut holes on the leaf or eat the margins of leaves. The grubs feed on the roots of plants and cause wilting and drying of plants. But these weevils are generally of minor importance, due to low populations of both the grubs and adults, and normally do not require specific chemical management.

Biology

The adult weevils lay about 300-500 eggs in the soil over a period of three months. The grubs hatch out in a week and start feeding on the roots of the hosts. The grub period is around 60-75 days. The full grown grubs pupate in earthen cells for 10-12 days. The emerging adults take to the aerial portions and start feeding on the leaves of the hosts. They are polyphagous and can feed on a variety of crop plants and other vegetation.

Management

In cases of considerable damage to the roots in a plantation, the root zone can be drenched with 0.1 % Chlorpyriphos or rake up a dust formulation of Chlorpyriphos or any other dust formulation in the soil. The adults can be warded off or killed by application of contact insecticides like chlorpyriphos 0.05%, Malathion 0.05 % or Quinalphos 0.05% or synthetic pyrethroids at 0.01-0.05 % depending on the chemical nature.

Weevil : *Blocyrus inaeagualis* : Curculionidae

The adult weevils eat the leaves and thus make them perforated.

Chapter - 25

COLEUS AROMATICUS

Lace wing bug: *Monantha globulifera* : Tingidae

Eggs are inserted slantingly into leaves and other tender foliage. Black spiny nymphs and lace winged adults suck the sap from leaves. Attacked leaves develop yellow patches become brown, shrivel and dry up in severe cases.

Leaf feeder : *Helicoverpa armigera* : Noctuidae

The female moth lays eggs on tender foliage. Young larva feeds on tender foliage and starts infesting the older ones in later instars. A single caterpillar can destroy many leaves

Spherical yellowish white, eggs are laid by the female singly on tender plant parts. Eggs hatch in 2-4 days. The larva feeds on the foliage. Newly emerged larvae are yellowish white, while the fully fed are about 3-4 cm long, greenish brown with deep yellow stripes, but with many morphotypes based on the host plant and even the surroundings. The pupa in earthen cocoon in soil takes 6-21 days. Total life is for 4-6 weeks with 5-8 generations / year.

Management

- Collect and destroy the larvae.
- Set up pheromone trap with Helilure at 15/ha and change the lure once in 15 days.
- Six releases of *T. chilonis* @ 50,000/ha per week coinciding with flowering time and based on ETL.
- Release *Chrysoperla carnea* at weekly interval at 50,000 eggs or grubs / ha from 30 days after planting.
- Spray HaNPV or commercial formulations of NPV

Chapter - 26

DAVANA GERANIUM BOURBON

Semilooper: ***Thysanoplusia(Plusia) orichalcea*** **: Noctuidae**

Bluish green or grey coloured eggs are laid singly on lower surface of leaves. The larvae are semi loopers and feed voraciously on leaves and cause defoliation of the plants.

Pentatomid bug: ***Dolycoris indicus*** **: Pentatomidae**

Both nymph and adult stages suck the sap from leaves, resulting in yellowing, curling and falling of leaves. The young ones and adults also suck the sap from petioles, and young shoots. The removal of cell sap cause the shoots to lose their turgor and thus the plant becomes unhealthy and of low growth.

Chapter - 27

FENNEL

Thrips : *Thrips flavus* : Thripidae

Infestation of thrips is generally during hot months with high night humidity. The attack by thrips leads to discolouration, bronzing and drying of affected leaves.

Semilooper : *Thysanoplusia orichalcea* : Noctuidae

Bluish green or grey coloured eggs are laid singly on lower surface of leaves. Larvae feed voraciously on leaves and defoliate plants.

Pentatomid bug : *Dolycoris indicus* : Pentatomidae

A summer pest. Maximum population of the bug was found during May. Both nymphs and adults suck sap from growing spikes and cause flower drop.

Chapter - 28

FRENCH BASIL(*OCIMUM BASILICUM*)

The cotton aphid :*Aphis gossypii*: Aphididae

Both nymphs and adults cause the damage by sucking the plant sap from shoot tips, flowers and leaves. Severe infestation may result in complete drying of affected parts. Attacked leaves curl up, wither and fall. These are yellowish green or brown coloured aphids. The total life cycle from first instar nymph to adult occupies an average of 8-10days. It reproduces parthenogenetically and viviparously. The female may produce 10-30 young ones in a life span of 10-12 days. The nymphs transform into adult in 5-8 days after passing through four moults.

Management

- Use entomopathogenic fungus *Fusarium pallidoroseum* to cause epizootics in aphids in the cowpea field.
- Spraying of infested crop with 0.05% malathion or dichlorvos with a spray fluid of 500 litre water per ha effectively control aphids. This pest has a quick multiplication rate and hence the insecticidal treatment has

to be repeated as soon as aphid population is found to have built up again. Application of insecticides to follow the rule-spray until the first drop drips. Spray only on the tips where the aphids congregate

Leaf eating caterpillar: *Pycnarmon caberalis*: Pyralidae

The adult is a small whitish beautiful moth, with spots on both the wings. Larvae are small and feed on leaves and cause severe foliage loss.

Whitefly: *Dialeurodes* sp.: Aleyrodidae

Eggs are laid singly on lower surface of leaves. Nymphs and adults suck the sap from leaves. Affected leaves turn pale, curl and dry, and sooty mould develops on the leaves and other parts where honey dew secreted by the pest has fallen.

Tingid bug/ Lace wing bug: *Monanthia globulifera* : Tingidae

Female bug inserts eggs into the tissue of the leaves. Black spiny nymphs and lace winged adults suck the sap from leaves causing yellowing, curling and drying of infested leaves.

Chapter - 29

ISABGOL

The cotton aphid: ***Aphis gossypii*****: Aphididae**

Symptoms of Damage

Both nymphs and adults cause the damage by sucking the plant sap from shoot tips, pods and leaves. Severe infestation may result in complete drying of affected parts. Attacked leaves curl up, wither and fall. These are yellowish green or brown coloured aphids. The total life cycle from first instar nymph to adult occupies an average of 8-10days. It reproduces parthenogenetically and viviparously. The female may produce 10-30 young ones in a life span of 10-12 days. The nymphs transform into adult in 5-8 days after passing through four moults.

Management

- Use entomopathogenic fungus *Fusarium pallidoroseum* to cause epizootics in aphids in the cowpea field.
- Spraying of infested crop with 0.05% malathion or dichlorvos with a spray fluid of 500 litre water per ha effectively control aphids. This pest

has a quick multiplication rate and hence the insecticidal treatment has to be repeated as soon as aphid population is found to have built up again.

Chapter - 30

LEMON GRASS

Shoot borer: ***Chilo infuscatellus*** **: Pyralidae**

This pyralid is more commonly known as a pest of sugarcane. Eggs are laid in batches on lower side of leaves. The emerging larvae move down and cut a hole at the ground level and enters the stem. Due to feeding of larva, the central shoot begins to wilt and dry causing dead heart. The yield of the crop is affected.New tillers may develop but will not have the physiological maturity and oil content at the time of harvest.

The cotton aphid: ***Aphis gossypii*****: Aphididae**

Both nymphs and adults (apterous and alate) cause the damage by sucking the plant sap from shoot tips and leaves. Severe infestation may result in complete drying of affected parts. Attacked leaves curl up, wither and fall. These are yellowish green or brown coloured aphids.

The total life cycle occupies an average of 8-10 days. It reproduces parthenogenetically and viviparously. The female may produce 10-30 young ones in a life span of 10-12 days.

Management is by using the entomopathogenic fungus *Fusarium pallidoroseum* to cause epizootics in aphids in the cowpea field and spraying of infested crop with 0.05% malathion or dichlorvos with a spray fluid of 500 litre water per ha effectively control aphids. This pest has a quick multiplication rate and hence the insecticidal treatment has to be repeated as soon as aphid population is found to have built up again. Application of insecticides to follow the rule-spray until the first drop drips. Spray only on the tips where the aphids congregate.

Spittle bug: *Clovia bipunctata*: Cercopidae

Nymphs and adults suck the sap from the young leaves causing yellowing of and crinkling of the leaves. If necessary, spray contact insecticides.

Aphid: *Macrosiphum miscanthi*: Aphididae

Aphids suck the sap from growing points of tender shoots and developing young leaves resulting in yellowing of leaves and shriveling of tender parts. The new growth will be stunted. As a management strategy, the newest shoot where the aphids first appear in crop, should be picked, and destroyed. Periodic removal of the same will completely eradicate the aphid.

Termite: *Microtermes obesi*: Termitidae

Termites are usually dry season pests. They feed on roots and also on the stem portion of the plant. Affected plants show wilting and drying of leaves followed by death of the plant. Providing irrigation regularly can reduce the attack drastically. The soil around the plants in the pockets of attack can be drenched with a solution containing neem oil and liquid soap @ 20 ml oil and 5-8 gm/ml of liquid soap, per litre. If chemical control is desired, drench with chlorpyriphos solution @ 5 ml/litre.

Chapter - 31

MENTHA OR MINT

Cabbage semilooper: ***Thysanoplusia orichalcea*** **:Noctuidae**

Eggs are laid singly on lower surface of leaves. The emerging larvae feed as a scraper initially and then feed voraciously on the leaves. The larva is greenish and hence not easily noticed initially. But when patches of damage occur in the crop, examine for the larvae and remove and destroy all the larvae. Chemical control is not advisable, and spraying extracts of other aromatic plants can deter egg laying very well.

Leaf webber: ***Syngamia abruptalis*** **: Pyralidae**

Eggs are laid on tender foliage. Larvae web the leaves together and feed on them. When fully grown, they pupate inside the leaf fold. Damaged leaves become skeletonized and dry up in heavy attack especially during dry weather. Spray with deterrent extracts of other aromatic plants.

Gram caterpillar : *Helicoverpa armigera* : Noctuidae

Spherical yellowish white, individual eggs are laid by the female on tender plant parts. Eggs hatch in 2-4 days. The larva feeds on the foliage extensively. Larvae eat for 18-25 days with newly emerged ones being yellowish white and full grown with dark broken grey lines on the side of the body. They have 5 instars. The larvae are cannibalistic in nature and can feed on one another when food is scarce.

Aphid: *Aphis affinis* :Aphididae

The pest is active during March-April. Colonies of nymphs and adults are found on tender shoots and lower surface of leaves and suck the sap. Attacked leaves fade, curl up and fall. Tender shoots dry. Black sooty mould is seen on infested parts due to secretion of honey dew. Examine the occurrence of the pest during every stage of the crop, especially after the pickings, when new sprouts emerge. Apply neem oil in soap for initial attack.If severe attack is noticed, non residual sprays like Dichlorvos@ one ml per litre can be applied.

Chapter - 32

OPIUM

Cut worms: ***Agrotis suffusa*****: Noctuidae**

Adult moth lays eggs in moist soil or on lower side of leaves. Dark brown larvae cut young plants at ground level and shoot of grown up plants. Damage results in growth retardation of plants and reduced yield.

Capsule borer: ***Helicoverpa armigera*** **:Noctuidae**

This is a medium sized moth yellowish brown in colour with a wing span of about 3 cm and a body length up to 1.4 cm. There is a dark border and a patch of black and black spots of different sizes on the forewing. The hind wings are lighter, whitish, with a broad blackish band along the outer margin.

Female moth lays eggs on tender foliage. Larvae initially feed on foliage and later damage developing spikes and capsules. Young larva feeds on tender foliage and starts infesting the capsules in later stages. Spherical yellowish white, individual eggs are laid by the female on tender plant parts. Eggs hatch in 2-4 days. The larva feeds on the foliage. Larvae eat for 18-25 days with newly emerged coloured yellowish white and full grown with dark broken grey lines on the side of the body. They have 5 instars. The larvae are cannibalistic in nature and can feed on one another when food is scarce.

Management

- Collect and destroy the larvae.
- Setup pheromone trap with Helilure at 15/ha and change the lure once in 15 days.
- Six releases of *T. chilonis* @ 50,000/ha per week coinciding with flowering time and based on ETL.
- Release *Chrysoperla carnea* at weekly interval at 50,000 eggs or grubs / ha from 30 days after planting.
- Spray HaNPV or commercial formulations of NPV

Aphid: *Myzus* sp. :Aphididae

Aphids cause damage to growing tender shoots by sucking sap. Affected shoots wither, get deformed and retard in growth. Forced water jets during irrigation can dislodge the aphids and manage their attack. When the attack becomes severe,a mild insecticide may be "spot sprayed" on the tips of the shoots where the aphids congregate.

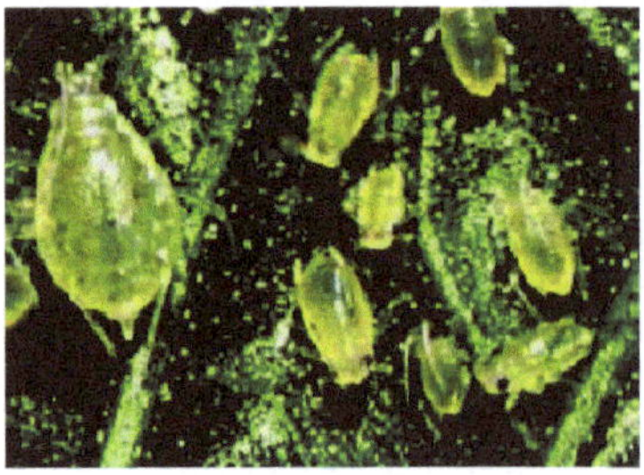

Chapter - 33

PATCHOULI

Leaf webber: *Syngamia abruptalis* :Pyralidae

Caterpillars feed extensively on tender foliage leaving veins alone and the growth of the plants are affected.

Leaf roller: *Psara stullalis*: Pyralidae

Larvae of these small moths are also leaf Webbers and rollers, and rolls the leaves and feed inside causing browning and drying of affected leaves. Inspect the crop for the pest incidence and remove the initial attacks systematically.

Pentatomid bug: *Nezara viridula* :Pentatomidae

Adults and nymphs of these green coloured bugs suck the sap from tender leaves and shoots. As a result, the leaves get deformed and growth of the plant is stunted. Inspect the crop for the pest incidence and remove the initial attacks systematically.

IPM in Aromatic and Medicinal Plants

Insecticide application in these crops is to be avoided. Most of these have repellency properties against pests. Use a crop mix that ensures deterrency to all the pests of the crops. This is ensured by planting specific crops in strips, so that volatiles from one strip deter the pests of the other strip. Crops that are least preferred by pests, should be grown as border crops. *Vitex nigundo, Nerium oleander* and lemongrass can be used for this purpose. Neem extracts and neem oil also can be used. Regular inspection of the crop and collection and destruction of the very first damage is the basic key to successful cultivation of medicinal and aromatic plants. These also ensure that the medicinal properties of the crop are conserved to the maximum.

Chapter - 34

PERIWINKLE

Sphingid larva: ***Dilephila nerii*** **:Sphingidae**

The large larvae cause severe damage to plants by feeding extensively on leaves, buds and flowers.They are very big and can easily be hand picked and killed or fed to poultry.

Leaf eating caterpillar: ***Anomis flava*** **: Noctuidae**

Larvae feed on leaves resulting in considerable foliage loss.

Grass hopper: ***Acrida exaltata*****: Acrididae**

Both young and adult stages of this grasshopper damage the foliage by making holes on leaves and cutting tender shoots.

Chapter - 35

SARPAGANDHA (*RAUWOLFIA SERPENTINA*)

Scale: ***Saissetia coffeae*: Coccidae**

Scale insects

Brownish black scale infests lower surface of leaves along with mid ribs and mid veins. They also cover petioles and tender shoots. Due to draining of sap, leaves turn yellow and drop, plants become stunted.

Grass hopper: ***Orthacris simulans*** **:Acrididae**

Both nymph and adult stages feed on leaves and cut tender shoots.

Leaf folder: ***Glyphodes vertumalis*** **:Pyralidae**

Larvae fold leaves together and feed on green tissue causing drying of shoot tip and stunted growth of the plant.

Chafer beetle: *Anomala polita* : Scarabidae

The young larvae cut the seedlings about 2cm above the ground. Adults emerge with onset of first showers and feed on foliage, buds and flowers at dusk. They have an yearly life cycle.

SECTION–III: ORNAMENTAL PLANTS

Chapter - 36

ANTHURIUM

Thrips – *Chaetanaphorothrips orchidii*, Thripidae: Thysanoptera

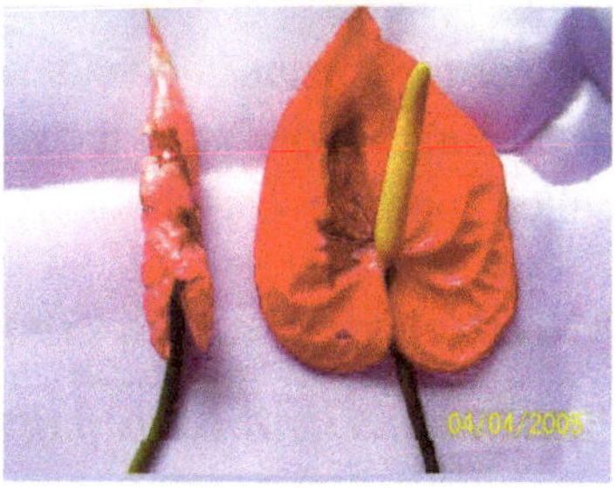

It is a polyphagous feeder that attacks other ornamentals, herbs, fruits, vegetables, grasses, and weeds, but shows preference for Anthurium species. It is a widely distributed species infesting green houses and out door landscapes.

Both adult and immature thrips start feeding within unopened Anthurium spathe. Damage appears as white streaks on the upper and lower surface of the spathe, resulting in deformed spathe and bronzing of the injured tissue with age. In severe cases, spathe fails to open and the foliage may be deformed with streaming and bronzing and thereby reduced plant growth.

Adult thrips are yellow in colour with banded wings. It lays about 80-100 eggs in a bud or sheath. Eggs hatch within 6-9 days. Nymphs and adults feed on the tissues of the plant. The life cycle is completed within 28-32 days and may extend up to 3 months.

Anthurium white fly:*Crenidorsum aroidephagus,Aleurotulus anthuricola* :Aleyrodidae

Anthurium white fly is becoming a pest of ornamental foliage plants. The adults and nymphs colonise on the under side of leaves and petiole sheath. Infestations by the nymphs may result in reduced vigor and general debility of the host plant. Heavy infestation may cause chlorotic spotting and localised necrosis. In addition to the direct feeding damage, the honeydew deposited on the leaves serves as a medium for the growth of sooty mould, which reduces the photosynthetic area and reduce the value of ornamental plants. The eggs are laid singly or in groups of two or three, scattered on the underside of the foliage. The first instar nymphs are grey and on emergence, actively crawl in search of feeding site. The second, third and fourth instars are stationery and are almost translucent. The adults are typical whiteflies, smaller than the common glasshouse whitefly, *Trialeurodes vaporariorum.A. anthuricola* have nymphs and adults with a white waxy coating and these colonise the underside of leaves and petiole sheath. The desapping causes yellowing, wilting and drying of the affected parts.

Aphids-*Aphis gossypi, Myzus persicae* Aphididae

Aphids feed on the underside of the leaves and tender growing tip, causing leaf discolouration, distortion and poor growth. They secrete a sticky sugary substance called honey dew resulting in the growth of black sooty mould over the foliage.

Mite:Red spider mite –*Tetranychus telarius* Tetranychidae

Brevipalpus spp. Tenuipalpidae

The population of mite may be abundant and reaches injurious levels during the dry season. Their presence on the leaf is recognised by a fine white web on the underside of the leaves. The minute red mites cause mottling of the leaves which may turn brown and later shrivel.

Snails

Many species of snails feed the root tips, and also damage the leaves and buds. Their abundance is mostly due to accumulation of moss and other saprophytic/semi saprophytic material that serve as their food. The major control, in this case, is hence, to prevent such material in the pots or other growing media. This can be accomplished by smearing any copper salt like copper sulphate in solution, on the sides of the pot, or growing benches. Also avoid excess irrigation.

Mealy bugs

Mealy bugs appear on the lower surface of leaves and also the roots. Honey dew and sooty mould may be present and the infected plants start drying.

Scale

It infests stem and leaves, suck sap and weaken the plants.

Root knot nematode

Nematodes occasionally attack the root of young seedlings and the plants become stunted.

Chapter - 37

CARNATION

Aphids: ***Myzus persicae*** **: Aleyrodidae**

Aphids are considered as the most serious pest of carnation. Both adults and nymphs suck sap from the leaves and debilitate the plants. They transmit carnation ring spot and carnation mosaic virus disease. They are thus more important as vectors. Regular monitoring for the incidence of the aphid and spot spray-ings of contact insecticides on the shoot tips can easily manage the pest.

Thrips : ***Thrips tabaci Haplothrips*** **sp. : Thripidae**

The adults and nymphs suck sap from the leaves, flowers and stem causing distortion, streaking and spotting. Severe attack adversely affects plant growth and flowering.

Red spider mite : ***Tetranychus urticae*** **: Tetranychidae**

These mites are also a serious pest of carnation. They live on the underside of the leaves and feed on the tissues. .Leaves turn pale and has a dusty coating of fine web. The affected plants become stunted. In severe cases, infestation spreads to buds and flowers and as a result the quality and yield of flowers are reduced. The buds and flowers do not open properly and the size, colour and appearance of the flower is lost.

Bud borer : *Helicoverpa armigera* : Noctuidae

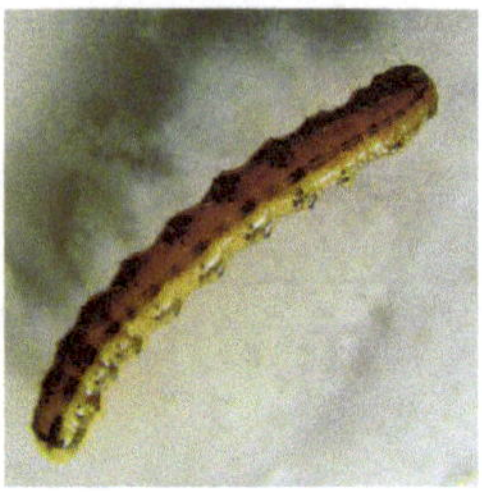

The adult moth of this boll worm lays creamy white eggs, singly on the flower buds. The young larvae feed by scraping the leaves. The late instars feed on petals of flower or buds by making round holes. The damaged flowers show an ugly appearance and buds fail to open as flowers.

Leaf miner : *Liriomyza* sp. : Agromyzidae

White maggots - *Hylemyia* sp. Anthomyiidae

The maggots bore into the leaves and growing tips of young plants. The leaf health is lost and the production of quality flowers is reduced.

Tortrix moth : *Tortrix pronubata* : Tortricidae. The larvae of this moth feed on the leaves, flowers and other aerial parts of the plant

Nematodes : *Criconemoides curvatum* and *C.xeroplex*

They feed on roots as ectoparasites and endoparasites. They cause a disease known as ring nematode decline, characterized by reduced root system, stunted growth of plants and reduction in the number of flowers.

Chapter - 38

CHINA ASTER

Chrysomelid beetle : *Aulacophora foveicollis* : Chrysomelidae

Adults cut tender shoots and feed on leaves resulting in holes and loss of leaf. Grubs feed on roots and underground stems leading to yellowing,wilting and drying of attacked plants.

Semilooper : *Clenoplusia albostriata* : Geometridae

These are deep green semi loopers, with wavy lines and they feed extensively on leaves.

Flower caterpillars : *Helicoverpa armigera* : Noctuidae, *Phycita* Sp.

The caterpillars feed on flower heads in large numbers and make them unfit for sale. Caterpillars of *Phycita* sp. are green in colour with brown head. Incidence of *H. armigera* is seen in the cooler months. Larvae damage flowers by making groove in the central receptacle and feed on ovaries and stamens of flower buds. Flower is lost causing full loss for the farmer.

Stem borer : *Platyptilia molopias* : Pterophoridae

The adult plume moth lays white elongated eggs on tips of shoots. Larvae bore into stem and side branches. Cream coloured larvae with black head feed inside shoots resulting in hollowing and wilting of affected shoots. Plants may die due to the attack.

Aphids : *Aphis* spp. : Aphididae

Root aphid : *Macrosiphum* spp. Aphididae

Cottony cushion scale : *Iceria purchasei* is sometimes a problem in temperate areas.

The Cabbage looper, Corn earworm, and cutworms also attack china aster.

Leafhoppers, Leafminers, soft scales and spider mites are also occasional pests in the crop.

Thrips in china aster include the Banded greenhouse thrips and the Western flower thrips

The Plant bug, *Lygus* spp is sometimes problematic in the china aster, where they suck the sap from the plant parts.

Whiteflies : The different species of whiteflies recorded on china aster include the Greenhouse whitefly, Iris whitefly, Silver leaf whitefly and Sweetpotato whitefly.

Chapter - 39

CHRYSANTHEMUM

Chrysomelid beetle : ***Aulacophora foveicollis*** **: ChrysomelidaeAphids :** ***Macrosiphoniella sanborni; Myzus persicae; Brachicaudus helichrysi; Aphis gossypi; Aphis fabae; Aulocorthum circumflexum*****: Aphid**

Chrysanthemum is infested by different species of aphids. Aphids are small, soft bodied insects and their colour varies with species. Both nymphs and adults cluster on the growing stem tips and under surface of leaves. As a result, the plants show loss of vigor, distortion of leaves and flowers, yellowing, premature leaf fall and stunted growth. The damaged plants bear only undersized buds and they remain unopened. Peak damage period is November-December. Secretion of honeydew by the aphids favor development of sooty mould which hinders photosynthesis and the plants become stunted. They are also vectors of virus diseases.

Macrosiphoniella sanborni

Thrips : ***Thrips tabaci, Thrips nigrophilosus, Frankliniella tritici, Hercinothrips sp., Microphalothrips abdominalis*** **: Thripidae**

Different species of thrips attack chrysanthemum plants. Thrips are small,

slender insects. The nymphs are generally straw white and adults are black coloured. Both nymphs and adults feed on tender leaves causing mottling, silvering and distortion of leaves. They also damage flowers of summer blooming cultivars. The affected flowers are discoloured, withered and dry up as if due to scorching heat.

Hairy caterpillar : *Spilosoma oblique* : Arctiidae

It attacks the plants in rainy season and continues till winter. Adult moths lay eggs in clusters on the upper side of leaves. The young larvae just after hatching feed on leaves gregariously. They skeletonize the leaves by scraping green matter and turn them to papery white and these leaves later dry up. Late instars spread throughout the field and cause defoliation by feeding the entire leaf lamina.

Leaf miners : *Phytomyza syngenesiae, Liriomyza trifolii, Chromatomyia horticola* : Agromyzidae

The incidence of leaf miner is a major problem in all chrysanthemum growing areas and especially when it is grown in green houses. The attack is severe in March-June. The flies lay eggs in the leaf tissues. The maggot just after hatching makes narrow, linear mines in the leaf by feeding the mesophyll tissues. In severe cases leaves show extensive mining which appear as zigzag silvery lines on the upper surface of leaves. The infested leaves later dry up.

The soybean leaf roller : *Hydylepta indicata* : Pyralidae

Adult moth lays eggs singly or in small groups on the lower side of the leaves. Eggs are pale white in colour. Larvae are green with brown head. It folds the leaves and feeds the green matter. Infested leaves become skeletonized and dry up. The larvae also damage flowers by feeding the petals.

Bud borer : *Helicoverpa armigera* : Noctuidae

The adult moth of American boll worm lays creamy white eggs, singly on the flower buds. The young larvae feed by scraping the leaves. The late instars feed on petals of flower or buds by making round holes. The damaged flowers show an ugly appearance and buds fail to open as flowers

White grub : *Holotrichia* sp. : Melolonthidae

White grub is a trouble some pest. The grubs in the soil eat the under ground portion of stem or root causing sudden wilting of healthy plants during dry hot moths.

Flower beetles : *Mylabris macilemta, Mylabris phalerata, Mylabris pustulata* : Meloidae

Blister beetles are voracious feeders of flowers petals. Severely damaged flowers lose their petals giving an ugly appearance.

Termites : *Microtermes obesi* : Termitidae

Termite attack becomes severe under dry soil conditions. They mainly feed on roots. Damaged plants wilt, dry and finally die if infestation is high.

Red spider mites: *Tetranychus urticae, Bryobis* sp. Tetranychidae

They are minute, red coloured arachnids found on undersurface of leaves causing white specks in the early stages. They occur in hot season. Damage starts in March and reaches peak during April –July. The affected leaves and buds become discoloured and fall off.

Semilooper : *Autographa nigrisigna* : Noctuidae

The caterpillars of this pest feed on flowers.

Leaf weevil : *Alcidodes fabricii* : Curculionidae

The adult weevil feeds on the foliage.

Flea beetle : *Altica* sp. : Chrysomelidae

The adults of this flea beetle can be seen feeding on the leaves and causing small shot holes.

Green Grasshopper : *Attractomorpha crenulata* : F Acrididae

The nymphs and adults of this flea beetle feed on the edges of leaves and petals of flowers in chrysanthemum.

Nematodes : Lesion nematodes: *Pratylenchus coffeae*

The nematode causes heavy root damage which subsequently leads to poor growth of chrysanthemum. The symptoms are in the form of stunting of plants with premature yellowing and drying of leaves, reduced flower size and dark lesions on the roots.

Chrysanthemum foliar nematode : *Aphelenchoides ritzemabosi*

This nematode causes considerable damage to the foliage of chrysanthemum. Internal discolouration of leaves is the characteristic symptom.

Chapter - 40

CROSSANDRA

Spikeborer : *Helicoverpa armigera* : Noctuidae

Larvae of this polyphagous moth feed on the spikelets, flowers and developing seeds. The young larvae bite holes into the spikelets and the grown up ones attack the mature spikes

Brown scale : *Saissetia nigra* : Coccidae; White scale: *Orthezia insignis*: Orthezidae

Yellowish nymphs and dark brown adult scales are seen in large numbers under the leaves and on petioles and stem.Sucking of sap in severe infestation results in stunting of growth and cause the leaves to turn yellow and drop. Honey dew secreted by The white scale favours development of sooty mould on the plant parts where the secretion falls.

Orthezia insignis

Saissetia nigra

Whitefly : *Lipaleyrodes* sp.: Aleyrodidae

It occurs in large numbers on the under surface of leaves. They desap from the under surface of leaves and the infestation leads to chlorosis and the development of sooty mould.

Pentatomid Bug *Gynencia affinis* : Pentatomidae

This is the crossandra bug, whose nymphs and adults suck the sap from the shoots and other aerial parts, resulting in brown lesions on the leaves and flowers. Severely affected plants fail to produce flowers. The bug is seen in abundance during winter.

Sorghum midge : *Contarinia maculipennis* :Cecidomyidae

The black coloured adult flies of this minute fly lays eggs singly within the buds of spikes. Orange maggots feed on ovaries of developing buds and pupate inside. Due to the damage, the buds fail to open and rot in severe cases.

Chapter - 41

GLADIOLUS

Thrips : ***Taeniothrips simplex*** **: Thripidae**

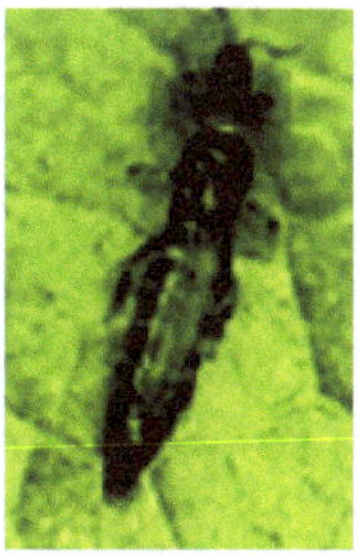

Yellow coloured nymphs and black adults damage leaves and spikes by rasping tissues and sucking the sap. Affected leaves and spikes develop silver streaks, turn brown, get deformed and dry when attack is severe. Corms in storage are also attacked. Infested corms are sticky, get shriveled and produce weak plants.

Cut worm : ***Agrotis segetum*** **: Noctuidae**

The female moth lays eggs near ground level on plant parts. The emerging larvae feed on developing shoots at night. The clay coloured larvae eat the shoots and grown up larvae cut the plants at ground level. Young plants up to around 3-4 leaf stage are most susceptible. The larvae also damage underground corms and developing spikes.

Leaf caterpillar : ***Spodoptera litura*** **: Noctuidae**

The adults lay eggs in groups on lower side of leaves. Early instar larvae feed on this lower side by scraping and skeletonizing them. Blackish brown mature larvae feed voraciously on leaves.

Mealy bug : *Ferrisia virgata* : Pseudococcidae

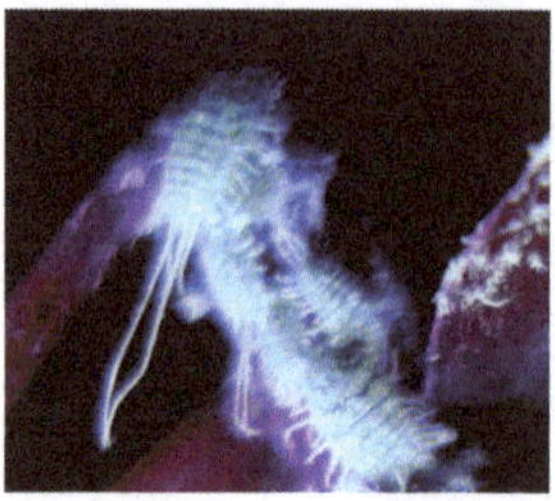

The infestation begins in the field on underground corms during dry conditions and is carried on to the storage. Nymphs and adults damage corms by sucking the sap resulting in shriveling and drying of affected corms.

Mite : *Tetranychus equatorious* :Tetranychidae

Attack begins at a very early stage when the plants are young. Green coloured nymphs and carmine coloured adults colonize on leaves and suck the sap. The damage results in discolouration, bronzing and shedding of affected leaves.

Lily bulb mite : *Rhizoglyphus echinopus* :Tenuipalpidae

These are Yellowish white in colour, with a pinkish tinge, bead like, shining & slow moving mites. They feed on the bulbs and roots and destroy them. The mites also infest the scales and burrow into the stems. The infested leaves and stems become yellow.

Chapter - 42

JASMINE

Bud-worm : *Hendecasis duplifascialis* : Pyraustidae

This pest is serious on *J.sambac*. The larvae are dark green with black head which bores into immature buds. They make a hole on the buds, and feed inside, leaving behind faecal pellets entangled in few silk strands. Webbing is seen in severe infestation and pupation takes place in soil.

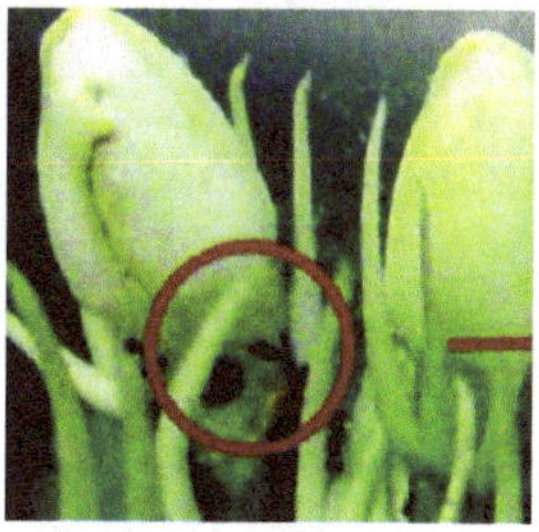

Leaf web worm : *Nausinoe geometralis*: Pyralidae

The larva is green with dark warts. Young larvae scrape the tissue, and mature larvae web leaves together with loose silken threads. Severely affected bushes show burnt appearance due to leaf drying. Pupation takes place inside the web. Damage is more in irrigated fields during summer.

Gallery worm : *Elasmopalpus jasminophagus* : Pyralidae

This is a serious pest on *J. auriculatum* and causes webbing on the terminal leaves, shoots and flower heads. The larva is green with a red head. It builds galleries of faecal matter and silk and feeds inside. Pupation takes place in the galleries.

Blossom midge : *Contarinia maculipennis* : Cecidomyidae

It is a pest on *J. sambac* and *J. auriculatum* .The attack causes the bud to show purple discolouration 4-5 days after oviposition. Maggots are dull white and feed on inner 1-4 petals, anthers and stigma. Mature yellowish orange maggots fall down to the soil and pupate.

Shoot borer : *Sycophila* Sp.: Eurytomidae

This is observed on *J. grandiflorum.* Larvae tunnel inside shoots and cause a series of grooves and pupate inside the shoots. Series of entry holes are seen on damaged shoots.

Gall mite : *Aceria jasmini* : Eriophyidae

It is a major pest of *J. auriculatum* and *J. pubescens.* The mites feed on the leaves and flowers and cause malformation and formation of felt outgrowths in vegetative and floral parts and consequent reduction in growth and yield.

Flower thrips : *Isothrips(=Thrips) orientalis* : Thripidae

The nymphs and adults of the thrips suck the sap from flowers and buds. The petals show brown streaks and buds fall off.

Chapter - 43

MARIGOLD

Bud caterpillars : *Helicoverpa armigera* : Noctuidae

Eggs are laid singly on young buds. Larvae feed on developing flowers by damaging florets. They can also eat the leaves.

Leaf webber : *Phycita* sp. : Phycitidae

Larvae of *Phycita* sp. feed on heads of buds and flowers.

Aphids : *Aphis gossypii* : Aphididae

Aphids mainly infest lower surface of flowers and base of petals. Nymphs and adults suck the sap from the flowers causing discolouration and withering. Congregation of the aphids and consequent desapping is also seen on the fresh shoots.

Thrips : *Neohydatothrips samayunkur* : Thripidae

Thrips infest young leaves, buds and flowers and suck the sap. Affected leaves get distorted, while petals of flowers turn brown and dry.

Chapter - 44

ORCHIDS

Biosduval Scale/Orchid scale : *Diaspis boisduvalii* : Diaspididae

These scales are round flat and yellow in colour. Only the first instar crawlers move. The others are sedentary. The young and mature scales suck the juice from the leaves, petioles and shoots and also inject toxins into the plant. Affected plants turn yellow in patches, lose vigour and get stunted.

Mealybug : *Pseudococcus maritimus* : Pseudococcidae

These are commonly referred as grape mealy bugs, yellow coloured with a covering of white powdery wax. They suck the sap from the plant parts and secrete honey dew. Sooty mould develops on leaves in case of severe infestation. Attacked plants turn yellow, withered and deformed.

Thrips : *Anaphothrips orchidaceous* : Thripidae

Both adults and nymphs suck the cell sap from leaves and growing buds. Affected leaves curl while petals develop streaks and the margins of leaves turn brown and dry up.

Giant African Snail : *Achatina fulica* : Achatinidae and other snails.

Adult and young ones of the snail eat the plants during night. Snail attack is heavy during monsoon period. They also cut tender shoots especially as grown up snails. The best way for control is to pick them and put in a container with salt. Spray the orchid pots on the outside, with 5% copper sulphate solution. Avoid excess irrigation.

Chapter - 45

ROSE

Hard Red scale: ***Aonidiella aurantii*** **:Diaspididae**

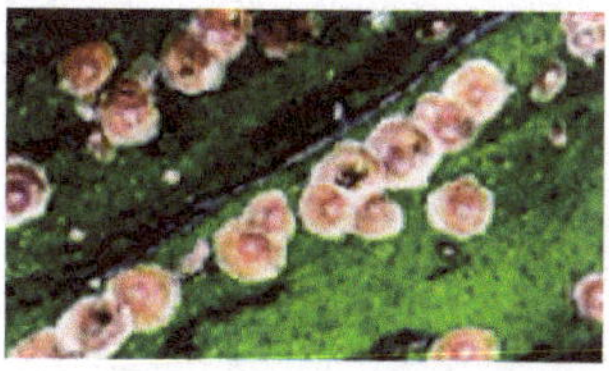

Both adults and young scales suck the sap from the stem and cause spot yellowing and loss of vigour of the plant. Infested shoots bear reddish brown encrustations of the scale and in case of severe damage the shoots dry up.

Scale: ***Lindingaspis rossi*****: Diaspididae**

The nymphs and mature scales of this hard scale suck the sap from the stems of rose especially in hot climate and cause devitalisation of the plant. An effective way of managing the pest is to smear the affected stems with a solution of 5% soap repeatedly for 2-3 times at weekly intervals.

Blackfly: ***Aleurocanthus spiniferus*** **:Aleyrodidae**

The puparia are black oval in colour and seen on the underside of leaves. The nymphs and adults drain the sap causing yellowing and crinkling of leaves.

Thrips: ***Rhipiphorothrips cruentatus; Scirtothrips dorsalis*****:Thripidae**

These are pests that attack the new flush that sprouts after pruning. Blackish brown adults and reddish nymphs of *Rhipiphorothrips* suck cell sap from tender leaves, buds and flowers. *Scirtothrips* in cream coloured is a major problem in polyhouse. Damage is seen as curled leaves with silvery yellow patches, brown marks and deformed buds with burnt margins.

Aphid: *Macrosiphum rosaeformis/rosae* : Aphididae

Blackish green coloured aphids cluster around growing shoots and buds. Both nymphs and adults suck cell sap and devitalize plants. High humidity and cloudy weather favours rapid buildup of aphids. Affected leaves get deformed while buds get retarded.

Leaf cutter bee : *Megachile anthracina* : Megachilidae

The adult cuts the leaves in a characteristic ring like fashion and almost fully remove it. The leaf is used in the nest by the bee.

Hairy caterpillar : *Euproctis fraterna* : Lymantriidae

Eggs are laid in clusters on lower side of leaves. Early instars are gregarious in habit and feed from surface causing skeletonization and drying of leaves. Black coloured mature larvae, feed voraciously on leaves, buds and flowers, resulting in severe defoliation and flower loss.

Hairy caterpillar/tussock moth : *Orgyia postica* : Lymantriidae

The larvae of this pest are defoliators on many crops including rose. They actively feed on the leaves. Bigger larvae can be mechanically picked and killed. Earlier larvae, usually in groups, can be collected from the leaves and killed.

Bud borer: ***Helicoverpa armigera*** **: Noctuidae**

Eggs are laid singly on growing buds. Larvae bore into buds and feed on developing petals. Open flowers are also attacked by larvae resulting in dropping of petals.

Crab caterpillar: ***Stauropus alternus /Neostauropus alternus*** **: Notodontidae**

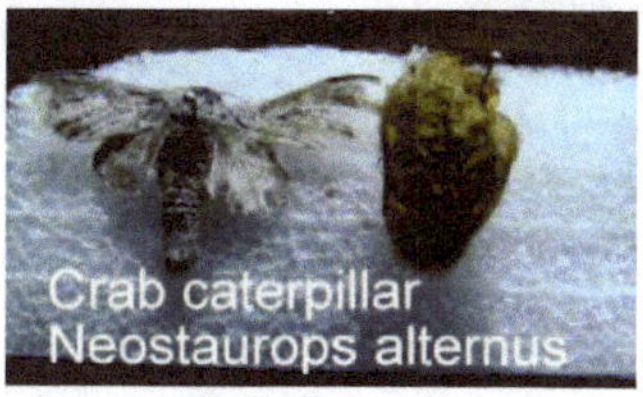

The larvare of this strange looking caterpillar will eat the leaves of rose.

Chafer beetle : ***Adoretus versutus*** **: Rutelidae (Melolonthidae)**

Damage starts with onset on monsoon. Black and brown coloured adult beetles emerge from soil in large numbers after dusk and feed on leaves, buds and tender shoots. Damage is mainly due to the grubs that feed on roots and cause the plants to lose vigour and die.

Ash weevil : ***Myllocerus*** **spp. : Curculionidae**

Eggs are laid in soil. The grubs feed on roots, while the grey coloured adults feed on leaves from margins in a regular fashion and cause severe foliage damage.

Termite : ***Microtermes obesi*** **: Termitidae**

Termite attack starts under dry soil conditions. The termites feed on roots and may attack the stem and bark in case of severe infestations. Affected plants wilt, dry and die consequently.

The minor pests on rose are listed below.

Hairy caterpillar:*Porthesia scintillans* : Lymantridae
Castor semilooper: *Achaea janata* : Noctuidae
Looper : *Agathia* sp.: Geometridae
Flower beetle : *Oxycetonia versicolor* : Cetonidae
Stem girdler : *Sthenias grisator* : Cerambycidae
Shoot webber/borer : *Acleris extensana* : Tortricidae

Two spotted spider mite : *Tetranychus urticae* : Tetranychidae

The mite is found in colonies in the lower surface of mature leaves under white-silky webs. In case of severe infestation, population spreads to entire foliage including flowers and buds. Hot and dry conditions favour mite multiplication. Cream coloured nymphs and red coloured adults with two black spots scrape and suck sap resulting in discolouration, bronzing and drying of affected leaves. Damage on tender buds lead to discolouration, retardation and drying of leaves and buds.

Pest Management

Roses are generally cultivated in pots or in microplots. The flowers being the produce, much care is to be taken in pest management. The usual pests are the summer pests like thrips, scales, and mites. Dedicated irrigation to reduce summer heat on the shoot tips will reduce their attack to a great extent. While contact insecticides are sprayed, add soap solution @ 0.5% to increase efficiency of control of the sucking pests. Caterpillars and flea beetles are easily controlled by the spray of contact insecticides. Root grubs and termites cause major damage to roots of rose. For management of root feeding insects, drench with Chlorpyriphos 0.05 to 0.1% depending on the severity of attack. Successful rose cultivation also requires systematic identification and eradication of rose diseases.

Chapter - 46

TUBE ROSE

Thrips: ***Thrips tabaci*** **:** ***Tuberose thrips***
Franklinella schultzei**: Thripidae**

These minute insects, preferring hot humid weather, rasp and suck the cell sap, causing bronzing, withering and drying up of leaves and flowers. This is the most serious pest on tuberose. Damage to young plants results in stunted growth.

Two spotted spider mite: ***Tetranychus urticae, T. telarius*** **:Tetranychidae**

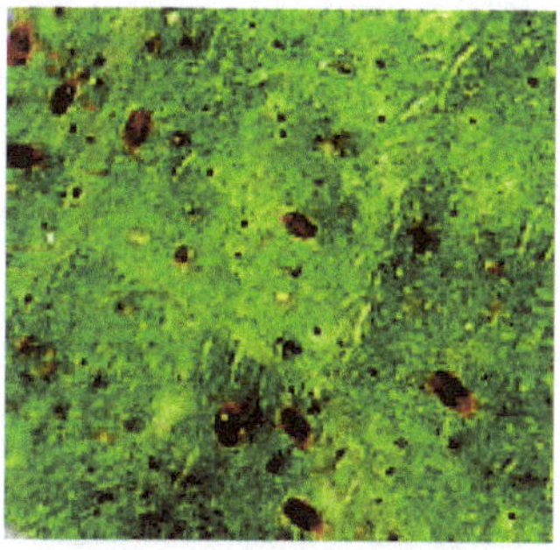

The mite is found in colonies in the lower surface of mature leaves under white-silky webs. In case of severe infestation, population spreads to entire foliage including flowers and buds. Hot and dry conditions favour mite multiplication. Cream coloured nymphs and red coloured adults with two black spots scrape and suck sap resulting in discolouration, bronzing and drying of affected leaves. Damage on tender buds lead to discolouration, retardation and drying of leaves and buds.

Bud borer: *Helicoverpa armigera*: Noctuidae

Eggs are deposited singly on growing spikes. Larvae bore into buds and flowers and feed on them by making holes, causing loss of flowers and fresh growth

Weevils: *Myllocerus* spp.: Curculionidae

Ash coloured adults feed on leaves by making characteristic cuts from margins during night.

Pentatomid bug:*Nezara viridula* :Pentatomidae

Eggs are laid in groups on leaves. Green coloured nymphs and adults suck the sap from basal portions of growing buds and flowers. Affected buds wither and drop off.

Aphids: *Myzus persicae* :Aphididae

Small green coloured aphids suck the sap from base of flower buds and growing spikes. Damaged buds and flowers fall off while affected spikes look withered and droop

IPM in Ornamental Plants

Ornamental and flowering plants grown in the open are best protected during rainy season by giving rain cover and during summer by giving rain type irrigation. Summer pests like thrips, mites, and whiteflies are regulated by decreasing the temperature in the microenvironment around the plants. Monitor pests everyday and mechanically remove and destroy affected parts. When insecticides are sprayed do not repeat the same chemical consecutively.

Grubs of chafer beetles can be managed by drenching the soil with chlorpyriphos @ 2.5ml/litre of solution. When population of aphids, thrips and leaf eating caterpillars increase, apply Methyl Demeton 2ml/litre or Dimethoate 1ml/litre or quinalphos 2ml/ litre in need based manner. Against budworms in jasmine, profenofos 1.5 ml/l or Lamda cyhalothrin 1.5 ml/l or Malathion 1ml/l can be alternated, based on requirement.

SECTION-IV: PLANTATION CROPS

Chapter - 47

ARECANUT

Spindle bug: ***Carvalhoia arecae*** **:Miridae**

Spindle bug is a serious pest multiplying rapidly with the close of monsoon. The adults are red and black and nymphs are greenish in colour with reddish brown patches. Both these stages colonize the top-most leaf axils at the base of the spindle. Adults and nymphs suck the sap from emerging spindle, tender leaves and other tender parts. For feeding they move out and later return to the axils again. At the site of feeding a longitudinal narrow, decolourised zone develops rapidly on either side of the puncture, along the length of the leaf blade. The decolourised patch gradually turns yellowish, then brownish, become stunted, twisted and malformed in a few days and dries up. In this way the tissues may break away and drop off in the wind leaving shot holes on the leaves. In cases of severe infestation the leaves get shred, stand erect and the palms become stunted.

The female bug thrusts the eggs singly into the tissues of the tender unopened leaves of the palm and rarely two or three eggs are found laid together. The operculum of the egg is visible with two silvery filaments projecting out. There are five nymphal instars which last in 13 to 20 days.

IPM of spindle bug

Filling the inner-most leaf axils around the spindle with phorate 10G (10 g/ palm) at three months intervals is reported to be effective. This is done by packing the insecticide in perforated sachets @1g/ palm and putting inside the inner leaf axils. As new leaves emerge, the perforated poly-bag sachets are to be shifted to the innermost leaf axils. Efficacy of the insecticide lasts for many months. Spraying carbaryl 50 WP @ 4gm per l at monthly intervals during peak occurrence is also recommended.

Root grub: *Leucopholis burmeisteri* :Melolonthidae and *Anomala varians*: Rutelidae

Root grub is a serious pest in loose sandy soil of arecanut. Severe attack causes yellowing of leaves, dropping of leaf or inflorescence, tapering of stem, reduction in yield and even death of young palms if the attack persists for 2-3 years. The grubs after the summer showers inhabit the upper layers of soil and feed on roots near the base of the palms causing the damage. The grubs attack and feed on roots of both young and old palms. They also feed on roots of any vegetation. In severe case, the entire bole region of the palm is destroyed. This is also a pest of coconut and thus is severe in mixed plantations of coconut with arecanut and other intercrops.

The adult beetles are chestnut coloured and come out of the burrows in the soil in large numbers during the first showers. The beetles lay the eggs in the soil a depth of 7 to 15 cm. The larval stage lasts for ten long months.

Root grub adult

IPM of root grub

- Keeping the garden clean, well-drained and free of weeds reduces its attack.
- Deep ploughing when the pest is found in abundance, especially after the pre-monsoon showers when large number of young larval instars are available in the top layers of the soil and expose these early stages to the predators especially birds.
- Crows are major predators on the adult beetles

- Loosen the soil around the base of the palm to a depth of 10-15 cm and drench with chlorpyriphos (0.04%) suspension twice, once in May just before the onset of southwest monsoon and again in September-October towards the close of the monsoon. Repeat the application for 2 or 3 years for complete eradication of the pest.
- Soil application of dust formulations of insecticides around the basins of palms just after the premonsoon showers to kill the eggs and first instar larvae gives satisfactory reduction in population

Black headed caterpillar: *Opisina arenosella* : Oecophoridae

Black headed caterpillar is occasionally seen as a defoliator pest of arecanut in India. Though the pest is prevalent all over peninsular India, it is more injurious along the coastal regions.

Inflorescence caterpillar: *Tirathaba mundella* :Pyralidae

The caterpillars are seen inside the unopened spadix where they feed voraciously within silken webbings. The webbed flower stalks has also the excreta in between. When the flower has opened, these webbers can be seen actively feeding within the webbings and amidst excreta. Due to this, the flower may fail to open in time as the inner floral parts are webbed together by caterpillars inside it. Drying up of inflorescence is another symptom. Attack by various slugs predispose the inflorescence to the attack of caterpillar.

The adult moth is brownish grey on the forewings with hay colour for the hind wing, and with fringes of hairs on the margins. They have a wing expanse of about 25 mm. They enter the unopened spathe through cuts or wounds and lay eggs. Egg to larva to pupa to adult takes 5-26-12 days.

Management

Force open the affected inflorescence out of the enclosing spathe and spray malathion 50 EC (200 ml in 100 litres of water).

Leaf caterpillar :*Elymnias caudata* :Nymphalidae

The adult is a medium sized grand butterfly. The wings are black with yellow patches in the centre. Eggs are laid singly on the leaves and they hatch in about a week. Larvae feed for about a month on the leaves. The full grown larvae are greenish with dorsal yellow stripes, rough bodied and has forked process at the tail end. The head is big, deep grey and has a pair of black horns. The pupa, which is a chrysalis, is green with red and yellow markings. The larvae are usually problematic in the nursery and for the seedlings.

Management is easily done by spraying any contact insecticide like quinalphos (2ml/l) or carbaryl 2gm/l.

Bark weevil: *Diocalandra stigmaticollis* :Curculionidae; Coleoptera

The weevil infests the cuts and wounds on the tree and the petioles. It burrows into the soft tissues of the bark through the tender leaf sheaths and lays eggs on the stem surface. Weevils also tunnel within the base of the leaf petiole which is close to the stem. Production of leaf and spathe is delayed. In severe cases leaves fail to emerge and trees become weak and yield reduces. Large number of adult weevils is seen in damaged palms within the base of the fronds. Adult is a small brownish weevil about 6mm long. This can be effectively managed by spraying or swabbing the tree trunks with 0.05% quinalphos after removing the hanging fronds.

Red Palm weevil: *Rhynchophorus ferrugineus* :Curculionidae

The red palm weevil attacks coconut, date palm, African oil palm, Palmyra palm, sago palm, royal palm, toddy palm etc. In general, the infestation by red

palm weevil is predominant in younger plantation between the age group of 5-20 years. The infestation starts with the female weevil laying eggs on the crown, stem or bole region. On stem and bole region eggs are laid on cuts or injuries present on these regions .Bole infestation generally occurs on young palms up to 7 years old. The grubs on hatching bore into and feed on the tender soft tissues. As they grow, the grubs tunnel extensively inside.

Yellowing of the inner leaves and gradual wilting of the central shoot in the crown, presence of cocoon, pupal fibre or chewed up fibers in the leaf axils or base of the palm , presence of small circular holes on the stem with a reddish brown viscous fluid oozing out, easy coming off of the leaves when pulled at, longitudinal splitting of the leaf petiole and emission of characteristic foul odor from the damaged portion of the crown are the main symptoms of red palm weevil attack.

The pest breeds within the palm. The adult is a large sized reddish brown weevil with 3-6 dark brown spots on the pronotum. Male weevil has tuft of short hairs on the dorsal side of the snout. The female weevil lays about 145-200 eggs. The incubation period ranges from 2-3 days. The grub period lasts for 55 days on an average. Fully grown grub is stout, fleshy, and apodous with conical body and reddish brown head. Pupation takes place in cocoon made of fibrous tissues which lasts for about 12-33 days.

Management

Integrating various measures like field sanitation, trapping techniques and prophylactic and curative chemical control methods brings about effective control of the pest. Remove and destroy all the wilting and dead palms to avoid breeding of the pest *in situ*. In attacked palms, seal all the holes on the stem with cement clay mixture except the topmost hole and pour 0.1% Carbaryl or 0.1% endosulfan suspension @ one litre per palm using a plastic tube attached to a funnel. If the damage is in the crown region, after cleaning the infested area without damaging the cabbage, pour the insecticidal solution slowly. To trap the adult weevils, tender arecanut logs split longitudinally with the cut surface smeared with fresh fermented toddy or yeast or acetic acid can be used. Log traps can be set in the plantation during dusk @ 5-6 traps per hectare and the weevils caught can be collected and destroyed next day morning. Commercially available male synthetic aggregation pheromone, 'Ferrolure' can be used to mass trap the weevils in plantations in wider area. The pheromone lure is used in food baited bucket traps and the trap is to be kept outside the farm preferably in a non-palm area, @ one trap per hectare. Fruits like pineapple or banana along with yeast and few granules of carbofuran can be used as food bait in the bucket. The trap should be cleaned once in 7 days and fresh food bait provided.

Pentatomid bug: *Halyomorpha marmorea* : Pentatomidae

The adult and the young bugs pierce and suck sap from the endosperm of tender nuts, causing premature shedding. Due to the feeding, tender nuts are shed. The dropped nuts have one or more pinprick-like marks on the surface below the perianth or on general nut surface where the nut has shrunk. The adults are bronze coloured with brown spots and in young stages they are black with white spot on the leg. When tender nuts are not available, the insect migrates to other hosts like cowpea and bitter gourd.

Control Measures: The crops should be closely watched and the bug when noticed should be mechanically removed and destroyed. One round of spray with Endosulfan (0.05%) or Fenvalerate (0.02%) on the bunches is effective.

Spotted grasshopper :*Aularches miliaris* :Acrididae

This is an occasional pest assuming locust proportions in some years, feeding on shade trees, arecanut, banana and wild plants. Nymphs and adults feed gregariously on the vegetation and produce a frothy secretion between the head and thorax when challenged. Adults are green and beautiful to look at, and spotted with yellow markings. In the swarm phase, rake the egg pods and expose the eggs in the soil, collect and destroy the nymphs and adults, and if needed, spray quinalphos 0.05 percent or dichlorvos 0.75 per cent.

Termite: *Odontotermes obesus* :Termitidae: Isoptera

Nymph and adult termite damages seedlings in the nursery and transplanted seedlings. They feed on the collar region of the seedlings and result in wilting of the central shoot. In grown up palms the termites attack the bark and runways made of soil and fibre are seen plastered on the tree trunk. The damage is more severe in the West Coast region .Drenching the nursery with 0.05% chlorpyriphos twice at 20-25 days interval manages the pest in the nursery. Infestation on older palms can be managed by spraying 0.05% chlorpyriphos on the affected trunk.

Leaf thrips: *Rhipiphorothrips cruentatus* :Thripidae

The adult females are dark blackish brown and males have bright yellow abdomen. The eggs can be up to 50 and a life cycle is completed in 15-25 days. The nymphs are pink in colour. Adults and nymphs suck the sap in colonies from the under surface of leaves. The desapping causes yellowing and later drying of the leaves.

Scale insect: *Aonidiella orientalis*: Diaspididae

The scale insects colonize leaves, spathe, bunches and nuts and suck the sap from the soft parts. Feeding on leaves and tender nuts result in yellowing of tissues. In severe cases, kernel may not develop and turns black. Nuts fall.

Chemical control is not advocated against the scale insects, as the beneficial biocontrol agents are killed by the insecticide spray. Ladybird beetles, *Chilocorus nigrita* and *Chilocorus circumdatus* (coccinellidae) are found to be effective biocontrol agents against the scale insects attacking areca palm. These can be released in the affected areca garden to control the scale insects.

Mites:*Raoiella indica* :Tenuipalpidae and *Oligonychus indicus*: Tetranychidae

Mites are commonly found in arecanut gardens but they pose serious problems only occasionally under prolonged dry weather conditions,. The mites are of 2 types-red(*R. indica*) and white(*O.*

indicus). Nymphs and adults of these mites congregate in colonies on the undersurface of the leaves and suck the plant sap. They attack both seedlings and adult palms. Mostly the damage is considerable only in the young palms. They infest the lower surface of leaf and suck sap and leaves turn yellowish-brown and dry up. As a result of their feeding yellow speckles are formed on the upper lamina. Later they coalesce becoming bronze coloured and finally the entire lamina dries up. The tender fruits are also attacked, causing malformation and shedding.

Calyx mite: *Dolichotetranychus* sp. : Tenuipalpidae

Colonies of mites are seen inside the perianth and due to their feeding tender nuts are shed. The mites can also cause drying of the young flowers.

Control Measures: Removal of heavily infested and dried leaves and burning help in eradicating the source of infection. Spraying under surfaces of leaves and crown with Dicofol (2 ml/litre) or Rogor (1.5 ml/litre) or formothion (Anthio 25 EC) 1.5ml/lit is effective. Repeat spraying at an interval of 15-20 days if there is recurrence of pest.

Many coleopteran predators, chiefly lady bird beetles (coccinellidae), are used against these mites. They are *Cyleophalus semipictus, Stethorus keralicus* and *Stethorus parcepunctatus*.

Eriophyid mites:Eriophyidae

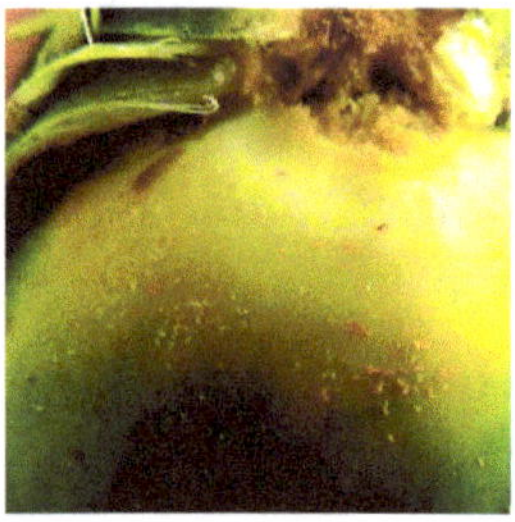

This author has also observed eriophyid mites under the perianth in some areca nuts in recent years, but the attack does not seem to be substantial.

Nematodes

out of the 28 genera of plant parasitic nematodes reported in arecanut, the burrowing nematodes (*Radopholus similis*) is the only endoparasite encountered in more than 50% of the areca root samples collected. Infested palms show general yellowing, reduced growth, vigour and yield. Appearance of orange-coloured lesions, blackening of tips of lateral and tertiary roots and rotting of roots are conspicuous symptoms.

Control Measures: Integrated methods to control nematode infestation are: use of resistant/tolerant varieties like *Sumangala* and *Sree Mangala*; avoiding susceptible hosts like pepper and banana in the areca plantation that would serve as additional hosts, using nematode-free planting material; application of 5-10 kg of green leaves/palm, preferably of *Glyricidia* or *Crotalaria*; applying neem cake (1 kg/palm/year); and applying phorate (3 g/plant).

IPM in Arecanut

Root grubs can be managed by applying chlorpyriphos @ 2.5 ml/l to the root zone just after the pre-monsoon showers. Set up light traps during dusk in the periphery of the garden during pre-monsoon showers when the adults emerge. Burn trash, old dried leaves, petioles and leaf bases during dusk to attract and kill adults. Mites can be managed with Dicofol 2ml/l, or Dimethoate1.5ml/l. Keep perforated sachets containing phorate 10G in the innermost leaf axils to deter spindle bugs. If inflorescence caterpillars are prevalent, force open the spadix and pour Malathion at 2ml/litre into the spadix.

Chapter - 48

CASHEW

The Tea mosquito bug : *Helopeltis antonii* : Miridae

The host range of this bug also includes tea, neem, moringa and guava. The attack can cause inflorescence blight, terminal drying of young shoots and heavy yield loss. The adult is a reddish brown bug with black head, red thorax, and black and white abdomen. The easiest identifying character for this pest is a pin like process on the scutellum mid dorsally on the thorax. It is reddish brown, erect and knobbed, and would seem as a pin inserted into the scutellum.

Eggs are inserted by the female into epidermis of tender shoot, axis of inflorescences and nodes. Eggs are elongated and slightly curved with a cap like operculum bearing two bristles. These project to the outside and show the sites of oviposition. The eggs are usually laid in two rows of 3 each, and about 30-35 eggs may be laid by a female. Egg period last for 7-8 days and the nymphal duration is 14-16 days. One life cycle is completed in 22-25 days.

Cashew drying by tea mosquito attack

The bug population starts building up from the flushing season in cashew. With the flushes proliferating and the development of the inflorescences, the attack also increases. The adults and nymphs feed on the new flushes, inflore-scences, and also the very small developing nuts. The secretion of saliva by the insect causes toxaemia, thus reducing the yield further. The area around the feeding site gets dried up. The attacked shoots show die back. This is a serious pest in cashew, limiting its production. In recent years the attack is seen combined with the infection by the fungus *Colletotrichum.*

Management

i. Regulating the shade to facilitate proper penetration of sunlight inside the canopy.
ii. A systematic management programme involving need based appli-cation of insecticides is recommended. The following insecticides are to be sprayed during early morning hours, thoroughly covering foliage and bark
    - At new flush formation during November – December, apply Monocrotophos @0.05%(1.25ml/l). Add copper oxychloride 0.2% if *colletotrichum* is prevalent.
    - Second application at flower initiation with Endosulfan / Quinalphos @ 2 g/lit + Mancozeb 0.2% if anthracnose is suspected.
    - Third application at fruiting time with Carbaryl 0.2%(4gm/l) .

Stem and root borer : *Plocaederus ferrugineus; P.obesus* : Cerambycidae

The host range of this pest also includes moringa, silk cotton and citrus trees. The borers cause wilt-ing of branches and sometimes the whole tree if the infestation is in the trunk and root. *P. obesus* is only a minor pest.

The adult of this long horned beetle is reddish brown and with the head and thorax dark brown or black. The female lays the dirty-white eggs under crevices on the bark of tree trunk or on the exposed roots. Grubs bore into the bark in their early stages and make extensive tunnels within. Later they attack the central wood and lives there chewing out frass to the outside. The grub feeds for 6-7 months and by this time the vigour of the tree is lost, showing yellowing and drying symptoms. The leaves are shed and branches, and later the whole tree dies. The tree produces its characteristic resin consequent to the attack, which solidifies into gum inside the tree. Hence the tunnels by attacking grub are not easily reachable to liquid formulations of insecticides. Larval period: 6-7 moths. Grub measures up to 7.5 cm. The grub pupates in a calcareous pupal chamber for two months. Insect completes one life cycle yearly.

Management

- Swab the trunk region up to about one metre height with a suspension of carbaryl 0.2% or chlorpyriphos 0.1% twice a year in endemic areas. Inspect the trees systematically and clean the collar region and also remove grubs, pupae and eggs.
- Break up aluminium phosphide tables and insert pieces of it into the bore holes and seal with a mixture of cement and mud paste. Most of the other pests on cashew are only minor pests.

Leaf miner : *Conopomorpha(Acrocercops)syngramma* : gracillariidae

This pest is seen immediately after the flushing in cashew. The caterpillar mines the tender leaves producing white blistered patches. These blisters remain in the older leaves after the adults have come out of the mines. This is not a serious pest to be controlled.

Leaf and blossom webber : *Macalla moncusalis* : Pyraustidae

These are medium sized moths up to 24mm in wing span. They are brownish males and pale and olive green females. The caterpillars web the terminal leaves and blossoms on the branches, feeding on them. If necessary, a contact chemical like Quinalphos, Carbaryl or triazophos can be applied to manage the larvae.

Inflorescence caterpillar :*Hypatima(Chelaria) haligramma*: Gelechidae

The adult is a small moth with a wing span of about 2 cm. The larvae are yellow or greenish brown. They feed on the shoot tips or inflorescences. They may also bore into the tender shoot tips. Pupation is in the folded leaves or in small holes on the shoot tips or floral parts.

Thrips: *Selenothrips rubrocinctus, Rhipiphorothrips cruentatus, Retithrips syriacus, Rhynchothrips raoensis* : Thripidae

All these are pests that attack the inflorescences, when they emerge. Most of them also feed on the leaves causing discolouration and crinkling of the leaves. The developing nuts are also attacked and show malformation, scabs and underdevelopment if the attack is severe. These become heavy in their population only when the summer is severe.

Nut bug : *Paradasynus rostratus* : Coreidae

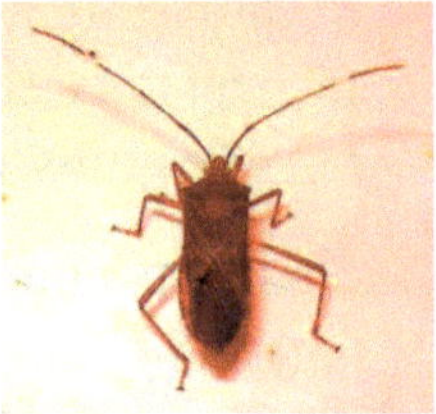

Eggs are laid by the female on the leaf surface in groups .They are boat shaped and yellow and hatch out in 8-10 days. The nymphs and adults feed on the developing nuts causing hem to shrink and dry up.

The brown aphid : *Toxoptera odinae* : Aphididae

The brown aphids may occasionally colonise the tender shoots and inflorescences.

Shoot borer : *Anarsia epotias* : Gelechidae

The adults are greyish with black patches having up to 13 mm wing span. The larvae web together the tender leaves and later bore into the terminal shoots. Pupation is within the leaf web or shoots. The attack causes drying up of flushes and inflorescences.

Bark borer : *Indarbela tetraonis* : Arbelidae

Caterpillars of these are bark borers that make galleries of silk and frass on the bark and feed within.

Slug caterpillar : *Latoia lepida* and *Thosea cotesi* : Cochlididae

These feed on the leaves leaving only the veins. They are also irritating to touch.

Leaf webber : *Orthaga exvinacea* : Pyralidae

These are seen sometimes webbing the leaves of cashew.

Hairy caterpillar : *Metanastria hyrtaca* : Lasiocampidae

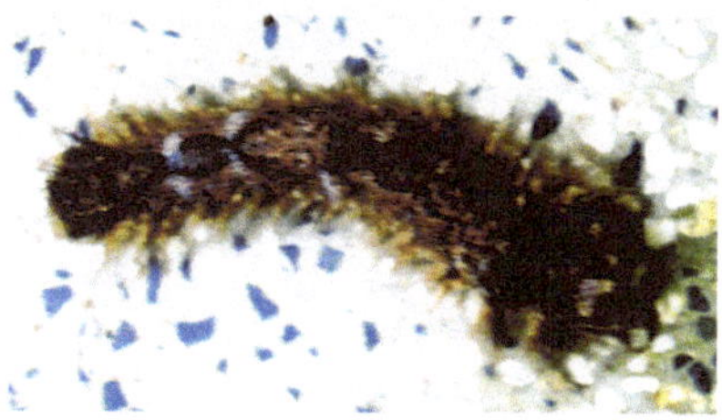

These are medium to large moths with a wing span of up to 7 cm. They are chocolate brown with a brown patch and a white spot on the forewings of the male. Females are robust with dark brown wavy markings on wings. The abdomen is long and has tufts of hairs at the tip. Eggs are laid on the undersurface of leaves in clusters. About 250 eggs may be laid by a female. The caterpillars emerge in about 9 days and the densely hairy caterpillars grow feeding on the leaves. Caterpillars are also big in size, measuring up to 7 cm in length. Defoliation by the caterpillars causes damage mostly to young plants.

Wild silk caterpillar : *Cricula trifenestrata* : Saturnidae

The moths of this silkworm are large reddish brown in colour with three spots on the forewing. The larvae are dark brown and grow to around 8 cm. The cocoon is with a golden yellow silk and is seen attached to the stems and branches of the tree. They are heavy defoliators and the pupae and the cocoons are seen in masses on the tree.

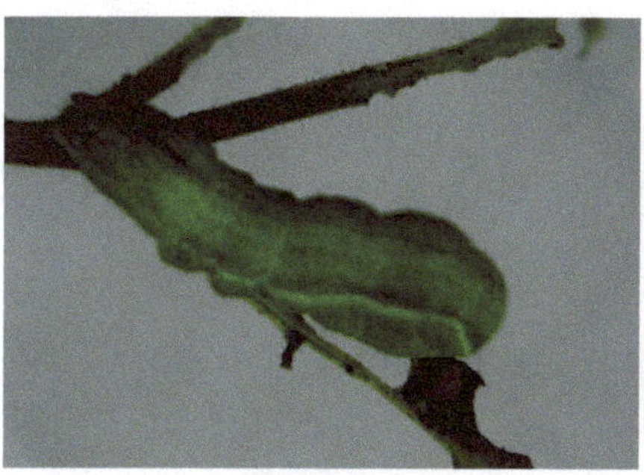

Wild silkworm : ***Antheria paphia*** **:Saturnidae**

These are seen sometimes feeing on the leaves causing heavy defoliation. The larvae are greenish and stout.Some effort was made to commercialise tasar rearing on cashew.

Cashew apple borer : ***Nephopteryx eugraphella*** **: Phycitidae**

These are occasionally seen on the apples consuming the flesh.

Chapter - 49

CACAO/COCOA

Cacao Mealy Bug : *Planococcus lilacinus* : Pseudococcidae

The adult female mealy bug is 3.9 mm in length and 2.25 mm in width. Eggs are laid on the plant parts within a mass of tangled waxy strands, a female laying 50 to 300 eggs. The eggs hatch in less than 6 hours and the nymphs feeding on the plant juice become full-grown in 20-25 days with a mean of 23 days in the case of females and 17 to 20 days for the males. The male nymph assumes a pupal period which lasts for 4 to 6 days and the adult male is winged. They live for 2-3 days while the adult female has longevity of 12 -16 days. The females dominate the population, 67.6 per cent of the crawlers developing into female mealy bugs. The mealy bug infests all tender parts of the cacao plant such as shoots, terminal buds, cushions, cherelles and pods. On the shoots the mealy bug prefers to feed from the growing tips. The feeding results in the suppression of the growth of tender leaves which get deformed into tender hair-like processes giving

the shoot tip the appearance of a brush. The jorquetting of the plant is also adversely affected. Colonisation of the mealy bugs on the cushions results in their abortion. The cherelles attacked wilt and drop. Shedding of cherelles (young pods) may be up to the extent of 70-80 per cent. On larger pods the sites of mealy bug feeding turn brown. Irregular minute cracks and pitting develop on the pod surface. The mealy bug is attended by around 7 species of ants. The insecticides effective in controlling the mealy bug are fenthion, monocrotophos, quinalphos and dimethoate applied as 0.05 per cent sprays. Addition of teepol or soap is needed for spreading of the insecticide and to dissolve the waxy coating of the mealy bugs.

Mealy Bugs : *Planococcus citri* and *Rastrococcus iceryoides* Pseudococcidae

They occur on tender shoots, flower stalks, foliage and on the developing pods. The tender pods when attacked dry up and on the mature pods sunken necrotic lesions develop affecting their full development.

Tree Hopper : *Gargara (Maerops) mixta* : Membracidae

The adult is a brownish cow bug and nymphs are green in colour. The female deposits eggs in slits cut on the tender shoots and these punctures eventually dry up. They cluster in large numbers on cocoa shoots and pods. Cherelles when infested may dry up while in the surface of grown up pods become brown. The red ant *Oecophylla smaragdina* attends the bug colo-nies and encloses them in webs.

Spittle Bug : *Phymatostetha deschampes; Peuceptyellus sigillifer* : Cercopidae

Of the two species recorded,e *Phymatostetha deschampes* is a red and blue bug usually seen on banana and the other is *Peuceptyellus sigillifer*.

Leaf Hoppers : Fam. Cicadellidae

The leaf hoppers recorded on cocoa are, *Bothrogonia* sp. *Krisna* sp. *Kurtara brunnescens* Dist. *Balocha* sp. and *Coelidia* sp.

Aphid : *Toxoptera aurantii* : Aphididae

The black aphid infests the growing shoots, underside of leaves, cushions, flowers and young pods resulting in the shortening of internodes, reduction in the size of leaves and deformation of leaves. When the cushion is attacked the flowers drop. Ants attend the aphid.

The Mango Mealy Bug : *Drosicha stebbingi* : Margarodidae

The nymphs and adult females of this large mealy bug cluster on the tender shoots feeding on the sap. This is serious only in few situations.

Mirid Bug : *Helopeltis antonii* : Miridae

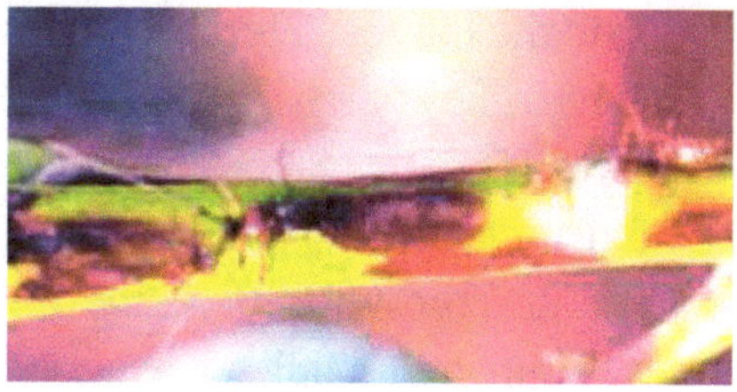

This is also called as tea mosquito and sucks the sap from shoots and pods. On the pods circular pitted black lesions appear as a result of the feeding of the nymphs and adults. Multiple feeding punctures make the pod surface rough and rugged and the pod deformed. Tender pods drop due to the attack.

Leaf Hopper : *Ldioscopus clypealis* : Cicadellidae

The small brown leaf hopper infests the tender leaves.

Order : Thysanoptera-Thrips

The Leaf Thrips : *Selenothrips (Heliothrips) rubrocinctus* : Thripidae

The adults and nymphs appear in colonies on the lower sides of leaves and occasionally on the pods. The infested leaves turn pale-green to pale-brown and dry up. The pods develop scabby patches around the feeding sites.

Stem Borer : *Zeuzera coffeae*: Cossidae

The adult is a leopard moth with white wings and dark spots on them. Eggs are laid in groups on small stems and branches. The young larvae make galleries in small branches and fan shoots and then find their way into the thicker stems. The full grown larva is red with brown head and is about 6.5 cm long. The larval period has duration of 60 to 100 days. Pupation is within the larval gallery with an exit hole for the moth provided. The total life-cycle is of four to five months duration. The caterpillar makes long galleries with openings on the bark surface through which a mixture of reddish ooze and frass emerge. Pellet-like reddish excreta is ejected out through the bore holes. These pellets are often seen accumulated on the ground. The attacked parts die and in the case of saplings the whole plants die. Pruning and burning the affected fan shoots and spraying carbaryl or chlorpyriphos 0.1 per cent on the stems of young plants as a preventive measure are recommended.

Looper Caterpillar : *Hyposidra talaca* : Geometridae

The adult is a light pink moth which lays eggs on the leaves, a female laying about 250 eggs. They hatch in 4 days. The looper larva feeds on the leaves and becomes full grown in 19 to 24 days. The grownup caterpillar is brown with orange dots on the sides. It pupates on fallen leaves. Life-cycle has duration of 30 to 35 days. The other looper caterpillars recorded on cacao are *Oenospila (Thalassodes) quadraria* and *Pingasa ruginaria.*

Tiger Moth Caterpillars : *Spilosoma (Diacrisia) oblique* : Arctiidae

The adult is a pale buff-coloured moth with black dots on wings. The orange coloured hairy caterpillars feed on tender leaves of cacao.

Pericallia ricini and *Amsacta(Creatonotus) gangis*: both from Arctiidae are the other hairy caterpillars, black in colour and these defoliate the plants.

Indian Gypsy Moth/ Tussock Moth : *Lymantria obfuscata* : Lymantriidae

The female moth is bulky and has rudimentary wings and hence cannot fly. The male is winged and free flying and brown with wavy markings on forewings. Eggs are laid among fallen leaves or on stem base . About 600 eggs are laid by a female and they hatch in 4 days. The hairy caterpillars feed on the leaves of cacao and when full-grown are brown in colour and 3 cm long. Larval period is completed in 33-38 days. Total life-cycle occupies 45-52 days. The caterpillars are voracious feeders of the leaves causing considerable damage.

They hide during day time in the ground under fallen leaves or may be seen clustered at the base of the stem. They pupate on the stem at its base or under fallen leaves within brownish cocoons.

Hairy caterpillar: *Lymantria ampla* : Lymantriidae

The female lays upto 1000 eggs. They hatch in 9 days. Larval period is 31-38 days in female having 5 instars and 27-34 days in male with 4 instars. The hairy caterpillars feed on tender flushes.

Hairy caterpillar: *Euproctis subnotata*; *Euproctis guttata*: Lymantriidae

The yellow moth lays eggs in masses of 8 to 10 on lower leaf surface. The hairy caterpillars feed on tender leaves and on the surface of tissues of young pods. The attacked pods dry up. The full-grown caterpillar is 4 cm long.

Hairy caterpillar : *Dasychira mendosa* : Lymantriidae

The moth is smoky brown with forewings carrying dark patches and pale grey hind wings . Eggs are laid in masses on leaves. The caterpillars feed on tender leaves, buds and on the surface of tender pods.

Leaf Caterpillar/semilooper : *Achaea janata* : Noctuidae

The caterpillar is long dark and with blue, red and yellow stripes. The moth is greyish brown with wavy markings on forewings and black and white patches on hind-wings. Eggs are laid singly on leaves. Egg, larva and pupa complete their growth in 4, 20 and 8 days respectively. Caterpillar is a defoliator. Pupation is in leaf folds.

***Leaf eating caterpillars* : Spodoptera litura : Noctuidae**

The caterpillars are sometimes seen feeding on leaves.

Hypsid Caterpillars : *Argina cribraria* : and *A. syringa* : Hypsidae

The hairy larvae of these feed on tender foliage.

Pod Borer: ***Conogethes (Dichocrocis) punctiferalis*** **: Pyralidae**

The caterpillar occasionally bores into the tender pods and makes galleries below the pod surface. Frass emerges from the entry holes. The seeds are not usually damaged.

Slug Caterpillar: ***Latoia lepida*** **: Cochlidiidae**

The slug caterpillar is occasionally seen on middle-aged leaves making large holes on them.

Leaf webber: ***Adoxophyes privitana*** **: Tortricidae**

The small caterpillar webs together growing points and feeds from within.

Chafer Beetle: ***Popillia complanata*****: Rutelidae**

The adult is a foliage feeder especially on the young foliage. Irregular holes are eaten on the leaves which often become extensive. The veins are left out. When disturbed the beetle drops to ground.

Cockchafer Beetle: ***Leucopholis*** **sp.: Melolonthidae**

Roots of cocoa plants grown in coconut gardens are eaten and damaged by the grubs. These feed on the roots of many plants including weeds. The attacked plants show yellowing and drooping of the entire foliage. Drenching the soil with chlorpyriphos 0.2 per cent suspension controls the grubs. This should be done at the beginning of the monsoon to kill the first instar grubs.

Flea Beetles : ***Monolepta longitarsus*** **and** ***M. limbata*** **: Chrysomelidae**

M. longitarsus is red in colour and *M. limbata* is dark coloured. The beetles eat minute circular holes on the leaves and in severe infestations the younger leaves get riddled with these holes.

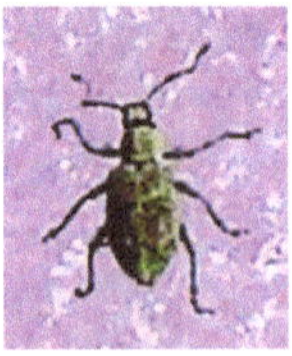

Leaf weevil: ***Myllocerus viridanus*****;** ***M. curvicornis*** **and** ***M. maculosus*** **: Curculionidae**

The greenish and grey weevils feed on the leaves from the margins consuming the lamina in irregular patches.

Stem Girdler : *Sthenias grisator* : Cerambycidae

The beetle girdles the main stem as well as thick fan branches and causes breakage at the girdled regions. The infestation is seen more near forest areas.

Stem Borer: *Glenea* sp. : Cerambycidae

The adult is 2.5 cm long with yellowish elytra having black dots. The grub girdles the stem of young plants. It feeds for some time on the bark remaining protected within a gallery after which it bores into the wood. Young plants dry up due to the attack by the grubs. *Pterolophia annobonae* also is recorded on cacao.

Powder Post Beetle : *Sinoxylon* sp. : Bostrichidae

The adult beetle bores into the stem and twigs making horizontal tunnels within. The larva also bores into the twigs of mature plants. The young seedlings are sometimes attacked and they die. Presence of dried-up leaves and twigs is the symptom of attack.

Non-Insect pests

Striped Squirrel: *Funambulus tristriatus tristriatus* : Rodentia: Sciuridae

The squirrel gnaws irregular holes through the walls of bronzing or mature pods and extracts the internal contents and feeds on the mucilaginous pulp surrounding the beans. The extent of loss to pods in Kerala ranges from 25 to 60 per cent. For management, harvest the pods when bronzing in just visible in the pod furrows. Also, mechanically protect the pods by covering them with punched poly-bags (150 gauge) smeared with bitumen-kerosene mixture.

Rat : *Rattus rattus wroughtonii* : Rodentia : Muridae

This is a serious menace to the cacao crop raised in coconut gardens. It feeds on maturing pods by gnawing holes on them. Keep coumarin or bromadiolone anticoagulant bars on coconut trees and use of bamboo traps to control them.

Order : Isoptera - *Odontotermes obesus* : Termites

Termites or white ants have been reported as killing young plants in the nursery and field. The collar region is the main part attacked and roots also may be damaged. White ants can be controlled by application of dusts of chlorpyriphos or fenvalerate or drenching with chlorpyriphos (0.1%) or drenching Imidachloprid @1 ml/l.

Chapter - 50

COCONUT

Though the coconut palm is subjected to attack by a number of insect and non insect pests although only 8-10 species of insect pests are considered to be important. Others assume importance, once in a while at isolated locations.

Major pests

The major pests of coconut include the rhinoceros beetle, red palm weevil, black headed caterpillar, root grub and the perianth mite.

Red Palm weevil : *Rhynchophorus ferrugineus* : Curculionidae

The red palm weevil is the most dreadful enemy occurring in all the coconut growing tracts of the world. The weevil attack can turn fatal to the crop if proper remedial measures are not adopted in time. Its host range includes date palm, African oil palm, Palmyra palm, sago palm, royal palm, toddy palm etc. In general, the infestation by red palm weevil is predominant in younger coconut plantation in the age group of 5-20 years. The infestation starts with the female weevil laying eggs on the crown, stem or bole region. Injury to the crown caused by rhinoceros beetle, bud rot and leaf rot diseases predisposes the palm to the crown damage by red palm weevil. On stem and bole region, the eggs are laid on cuts or

Red palm weevil damage

injuries. Bole infestation generally occurs on young palms up to 7 years of age. The grubs on hatching bore into and feed on the tender soft tissues. As they grow, the grubs tunnel extensively inside.

Symptoms of attack usually become clear only at an advanced stage of attack when the crown topples. But certain indications of attack can be detected early if closely looked for. Yellowing of the inner leaves and gradual wilting of the central shoot in the crown, presence of cocoon, pupal fibre or chewed up fibers in the leaf axils or base of the palm, presence of small circular holes on the stem with a reddish brown viscous fluid oozing out, easy coming off of the leaves when pulled at, longitudinal splitting of the leaf petiole and emission of characteristic foul odor from the damaged portion of the crown are the main symptoms of red palm weevil attack.

The pest breeds within the palm. The adult is a medium sized (35 x 12 mm) reddish brown weevil with 3-6 dark brown spots on the pronotum. Male weevil has a tuft of short hairs on the dorsal side of the snout. The female weevil lays about 150- 250 eggs. The incubation period ranges from 2-3 days. The grub period lasts for 55 days on an average. Fully grown grub is stout, fleshy, and apodous with conical body and reddish brown head. Pupation takes place in cocoon made of fibrous tissues which lasts for about 12-33 days.

Management

Integrating various measures like field sanitation, trapping techniques and prophylactic and curative chemical control methods brings about effective control of the pest. Remove and destroy all the wilting and dead palms to avoid breeding of the pest in situ. Destroy all the life stages inside the cut stem of the attacked palms by burning dry fronds over the rotten trunk. Leaving the trunk as such allows the adults to emerge out and attack other palms. Hence burning and killing them with dry fronds is very important. Resort to periodical crown cleaning of palms and avoid inflicting mechanical injury on trees in plantation which serve as ovipositional sites for the weevil. The wounds on the tree trunk should be treated with tar. Palms showing rhinoceros beetle damage, bud rot and leaf rot infection should be treated timely. Prophylactic leaf axil filling as suggested for rhinoceros beetle will give protection to the palms against red palm weevil also. Whenever leaves are cut from the palms, stumps of not less than 120 cm should be left behind from the base. In attacked palms, seal all the holes on the stem with cement clay mixture except the topmost hole and pour 0.1% Carbaryl or 0.1% dichlorvos suspension using a plastic tube attached to a funnel. If the damage is in the crown region, after cleaning the infested area without damaging the cabbage, pour the insecticidal solution slowly. To trap the adult weevils, tender coconut logs split longitudinally with the cut surface smeared with fresh fermented toddy or yeast or acetic acid can be used. Coconut log traps can be set in the plantation during dusk @ 5-6 traps per hectare and the weevils caught can be collected and destroyed next day morning. Commercially available male synthetic aggregation pheromone, 'Ferrolure'can be used to mass trap the weevils in plantations in wider area. The pheromone lure is used in food baited bucket traps and the trap is to be kept outside the farm preferably in a non-coconut area, @ one trap per hectare. Fruits like pineapple or banana along with yeast and few granules of carbofuran / a few drops of malathion can be used as food bait in the bucket. The trap should be cleaned once in 7 days and fresh food bait provided.

Rhinoceros beetle : *Oryctes rhinoceros* : Scarabaeidae

The rhinoceros beetle or black beetle is a widely distributed pest in all the coconut growing regions of the world. Apart from coconut, the beetle also attacks oil palm, date palm, Palmyra palm, pineapple, sugarcane and *Aloe vera*. The damage is inflicted by the adult beetle which bores into the central unopened spindle, spathe, leaf petiole and inflorescence stalk by remaining in between leaf sheaths near the crown. The beetle chews up the internal tissues and after ingesting the juicy part throws out the fibrous material which are seen extruding from the entry points. Due to the damage on unopened central spindle, the leaf is cut across in its folded condition. As a result, when fully opened, the affected frond shows characteristic V shaped or diamond shaped cuttings and the functional leaf area is reduced considerably. another visible symptom of attack includes presence of holes with chewed up fibrous material at the

base of the central spindle. Attack by the adult on the spathe causes wilting of inflorescence, amounting to direct crop loss. The damage done by the beetle provides egg laying site for the lethal pest, red palm weevil and for entry of fungal pathogens causing leaf rot and bud rot. Frequent infestation results in stunting of the palms and death of the growing point in young plantations.

The pest breeds in manure pits, decaying organic matter and decaying stems and stumps of coconut and other palms, coir pith etc. The adult is a stout black or brownish black beetle about 5mm long with a cephalic horn which is shorter in female. The female beetle lays eggs in the breeding site at a depth of 5cm. On an average a female lays about 100 eggs and the egg hatches in 8-18 days. The larva is creamy white, 'c' shaped, stout and sluggish which grows feeding on organic matter. The grub period is about 130 days with three instars. Pupation takes place in earthen cocoon and the pupal period varies from 25-30 days. Adult emergence is observed to be maximum following the pre monsoon showers. Adult longevity is 3-4 months.

Management

For an effective control of rhinoceros beetle population, integrated management measures should be adopted on a community basis. IPM package against rhinoceros beetle consists of field sanitation, mechanical, chemical and biological methods of pest management.

Decaying trunk of trees in coconut gardens should be destroyed by burning them as they serve as breeding ground for the beetle. The adult beetles can be extracted from the crown with the help of a beetle hook during the peak periods of pest infestation (June-September). Application of carbaryl or Lindane or methyl Parathion 5%dust or oil cakes of neem or marotti (*Hydnocarpus wightiaia*) in powder form @ 250 g mixed with equal volume of sand in to the top most three leaf axils around the base of the spindle leaf. This is to be done three times a year during May, September and December. The floating population of adult beetles can be trapped using toddy or fermented castor cake in mud pots placed in the plantation. Installation of aggregation pheromone traps in the periphery of the plantation @ one trap per hectare is also recommended on an area wide basis to trap the beetles. The grubs can be managed by drenching the manure pits and possible breeding sites with Carbaryl 0.1% solution or carbaryl or Lindane or methyl Parathion 5% dust once in three months. Incorporation of the plant parts of *Clerodendron infortunatum or Chromolaena odorata* in the breeding site is found to be good. Wide and successful control of rhinoceros beetle has been achieved using the microbial control agents, green muscardine fungus (*Metarhizium anisopliae*) and Oryctes rhinoceros nudivirus (OrNV). Application of the green muscardine fungus, *M. anisopliae* in the breeding site of the pest @ $5x10^{11}$ spores/m3 is recommended to manage the grubs. Once the fungus establishes, it survives in the site for more than two years. OrNV can be disseminated by releasing the infected adults in the field @ 10-15 numbers/ hectare.

Black headed caterpillar : *Opisina arenosella* : Oecophoridae

Black headed caterpillar is a serious defoliator pest of coconut in India. Though the pest is prevalent all over peninsular India, it is more injurious along the coastal regions. Palms of all age groups are susceptible to infestation by the pest. Other host plants reported are Palmyra palm, date palm, sago palm, ornamental palms like fish tail palm, fan palm etc. Though the pest infestation is reported through out the year in mild to medium form, peak infestation is found to occur during summer months.

The caterpillar scrapes the chloro-phyll by remaining on the under surface of the leaflets within galleries of silk and frass. The attacked leaflets turn brown and dry up presenting a scorched appearance. In case of severe infestation, due to large scale drying of the leaflets, the whole plantation presents a burnt up appearance from a distance. Diagnostic symptom of attack is the presence of galleries on the lower surface of leaf with live and dead larval and pupal stages of the pest inside, dried up patches on the leaflets with upper epidermis intact and drying up of the outer and middle whorl

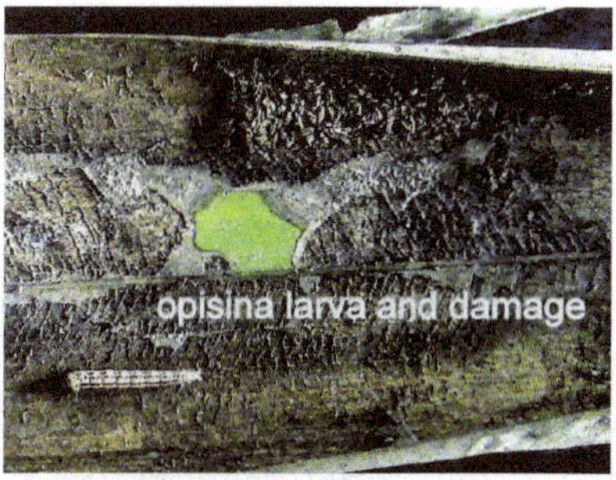

of leaves. Due to the leaf damage, there is reduction in the photosynthetic efficiency and hence nut yield is considerably reduced.

Adult is a greyish white moth, males smaller than females and with slender abdomen. Female moth lays eggs in small groups on the lower surface of leaflet near the old larval galleries. One moth lays on an average 137 eggs. The incubation period is five days. The caterpillar is with greenish to pinkish brown body and brownish black head and prothorax and brown longitudinal stripes on the body. Average larval period is 42 days including the pre-pupal period. It pupates in the larval web for in a silken cocoon. Adult emerges in 12 days.

Management

Biological control using parasitoids effectively controls the pest population. The larval parasitoid, *Goniozus nephantidis*, the pre-pupal parasitoid, *Elasmus nephantidis* and the pupal parasitoid *Brachymeria nasatoi* are recommended for field release. Initial assessment of the pest population and ascertaining the stage of the pest in the field population is necessary for effecting stage specific release of the parasitoids and fixing the dose of field release. Release *G. nephantidis* against the third instar larval stage, *E. nephantidis* for the pre-pupal stage and *B. nasatoi* to the early pupal stage of the pest at fortnightly intervals. If multiple stages of the pest occur simultaneously in the field, a combined release of all the parasitoids is required. However, during pest outbreak situation IPM measures involving pest surveillance, cultural, chemical and biological control methods should be adopted. During the pest out break, cut and burn one or two of the fully dried fronds in the outer whorl. Spray low residual insecticide such as dichlorvos @ 0.02% concentration if the pest infestation is severe and active larval stage is present. Subsequently, release stage specific parasitoids at fixed intervals. Release parasitoids only after three weeks of spraying. Adequate irrigation and manuring are to be given for rejuvenating the severely infested coconut palms during epidemic out breaks.

White grub : *Leucopholis coneophora* : Melolonthidae

Infestation by white grubs or root grubs in coconut is localised in the tracts of high rainfall and coastal sandy and sandy loam soil. Damage is caused by the grubs feeding and damaging the roots of coconut. The grub also feeds on the roots of sweet potato, colocasia, tapioca and banana raised as intercrops in

coconut plantation. Continuous feeding of roots by the grubs result in yellowing of the leaves, tapering of the stem and delayed flowering. In severe cases, immature nuts shed and the palm appears barren. Seedlings and young palms are most affected. Attack on young palms lead to stunted growth. In infested palms, white grubs are exposed when the base of the tree is dug.

The pest has an annual life cycle. The adults emerge with pre-monsoon showers and lay the eggs in the soil. Incubation period is about 20 days. The fully grown grub is creamy yellow and 'C' shaped with pale brown head. The grubs become fully grown in about 10-11 months in three larval instars. Peak grub population in the field is during July to October. The first instar larva feeds on grass roots and organic matter and the later stages feed on the roots of coconut and other intercrops. The final stage grub remains in pre pupal stage for 9-12 days before pupation in soil for about 25 days.

Management

The adult beetles can be attracted, collected and killed during peak period of emergence. Set up light traps at dusk in the periphery of the garden during pre-monsoon showers when the adults emerge. Trash, old dried leaves, petioles and leaf bases can be burnt during dusk to attract and kill adults. Plough the soil deep and expose the grubs during pre and post monsoon periods so that they will be eaten by predators and birds. The third instar grubs feeding on the tapioca sets can be exposed and killed. Apply Phorate 10G in the palm basin @ 100g per palm and incorporate in the soil to a depth of 30cm by raking. the root zone can otherwise be drenched with chlorpyriphos 20EC at 0.05% concentration (15-20 litres/ palm) during May-June and September –October. A waiting period of 45 days is recommended for nut harvest after the insecticidal treatment if systemic granules are used. In the field the grub is found to be preyed upon by the ground beetle, *Anthia sexguttata*.

Coconut perianth mite : *Aceria guerreronis* : Eriophyidae

The eriophyid mite, a native of Mexico was first reported in India from the Ernakulam

district of Kerala during 1998. Later on it has spread to all coconut growing states of South India as a serious pest. The mite is monophagous on coconut except an isolated report on the nuts of Palmyra palm.

The damage is caused by the nymphs and adults feeding on the tender tissues of developing nuts below the perianth. The mite gets entry into the buttons or developing nuts during the early phase of development immediately after pollination. Various stages of the mite are seen on the floral bracts and the soft meristematic tissues beneath the perianth. They pierce the tender tissue there and feed on the sap which causes death of the cells. Initial symptom of the mite infestation is noticed as white or yellowish white longitudinal streaks emerging out from just below the perianth. As the nuts grow these patches turn brown and longitudinal fissures and warts appear on the nut surface. Severe infestation causes drying and shedding of buttons or malformation of nuts resulting in retarded growth or shedding of immature nuts. There is considerable reduction in the copra content and husk fibre due to severe infestation. Mite attack is more severe during dry climate and drought situation aggravates the mite incidence. Peak population of the mite is reported during summer months. Studies in coastal Kerala revealed that a period of high temperature with intermittent rains causing high humidity favored higher multiplication and rapid spread of the mite. Usually 1-3 month old nuts harbor highest population of mites.

The mite is microscopic with an average size of 200-250 microns. It has worm like tapering body with two distinct parts, anterior cephalothorax and posterior abdomen and having two pairs of legs in the cephalothorax. Female lays about 60-200 eggs with an average fecundity of 80-150eggs. Eggs are glassy and transparent and hatches in 2 days. The nymphal stages are seen inside the perianth. When the population increases, adults move out of the perianth over the exposed portion of the developing nuts from where they are blown off by wind and get transported to fresh coconut buttons and start new colonies. A generation can be completed in 12-14 days.

Management

Though chemical pesticides like dicofol 0.1%, carbosulfan 0.05%, Profenophos 0.05%, triazophos etc reduces the mite population in the field significantly, these are not recommended for large scale adoption due to environmental reasons. IPM package recommended for the management of the eriophyid mite includes: Burning off of all the fallen nuts in the garden; spray application of Azadirachtin 0.004 % or neem oil- garlic- soap mixture 2% thrice a year during December-January, April-June and September-October. bunches of 2-4 months maturity alone need be sprayed. This is the stage which has the maximum mite population. The crinkling, warting and deep cuts in the husk starts only in later bunches, but these do not have significant mite population. The symptom development is entirely due to the defense reaction of the plant even in the absence of the mite. The unopened spathe, leaves and just opened inflorescences need not be sprayed. Insecticide is to be applied as fine droplets especially on the perianth region of the nut.

Coreid bug : *Paradasynus rostratus* : Coreidae

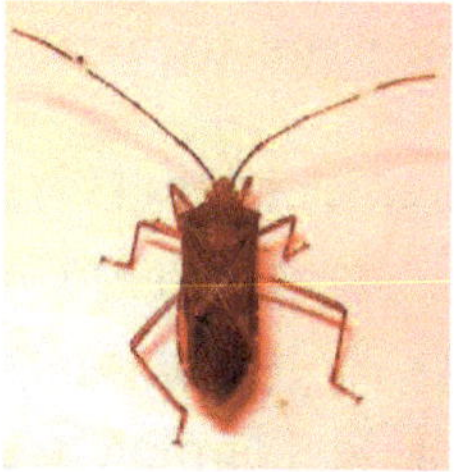

Both adults and nymphs of the coreid bug (also called as nut crinkler) feed on buttons and young nut. The feeding points on the nut are just below the perianth. The bugs pierce their stylets through the perianth and suck sap from the tender meristematic region of the coconut buttons. The feeding marks appear initially as brownish lesions and later these results in formation of cracks, shrinkage and malformation with gummosis on the affected side of the nut surface. Affected nuts are often shed. Those retained on the bunch become barren. Pest incidence reaches a peak in post-monsoon period from September to March. The host range of the bug includes guava, cashew, tamarind and cocoa. The adult is a large coreid bug, reddish brown above and pale beneath. The eggs are laid on leaf petiole, spathe, spadix or young buttons and hatch in 8-10 days. Nymphal development takes 30 days with 5 instars and adults live for 50 days.

The pest can be managed by spraying with 0.1% Carbaryl or endosulfan. Placement of phorate 10G at 5g/palm in two polythene sachets in the stalk region of youngest two bunches may deter the pest.

Mealy bug : *Pseudococcus longispinus, P.cocotis, Pulmiculter palmarum* : Pseudococcidae

The mealy bugs attack the spindle leaf, spadix and inflorescence. Due to the attack on the spindle leaf, it fails to open and remains highly stunted, deformed and suppressed. Infestation on the spathe and inflorescence cause poor development of young buttons and poor nut set. Mealy bug infestation can be detected by the presence of waxy white powdery coating and large number of ants moving around the region. Some times symptoms of leaf rot disease may be caused by this mealy bug. Mealy bugs can be managed by spray application of Quinalphos 0.05% or dimethoate 0.05%. Coccinellids and lycaenid caterpillars like *Spalgis epius* are found to prey on the different species of mealy bugs attacking coconut.

Scale insects : *Aspidiotus destructor, Aonidiella orientalis, Lepidosaphes mcgregori* : Diaspididae

These scales usually infest the lower surface of the leaf lamina in colonies. It may also infest petiole, peduncle, buttons and nuts. Nymphs and adult females feed by sucking the sap. Infestation leads to yellowing of the leaves in specks on the areas where the scales are present. Scale infestation peaks during summer months. In case of severe infestation spraying with fenthion 0.1% or dimethoate 0.05% or Monocrotophos 0.05% is recommended. In the field, scales are brought under check naturally by the coccinellid predator, *Chilocorus nigritus*.

Slug caterpillars : *Latoia lepida, Contheyla rotunda, Macroplectra nararia* : Cochlididae

The slug caterpillars attack coconut sporadically causing serious damage

occasionally. The caterpillars damage the plant by defoliation. They feed gregariously initially and disperse subsequently and feed voraciously on the leaflets leaving only the midrib. The slug caterpillars prefer young trees for attack. They appear fleshy and slug like with spiny scoli on the body. While mechanical control or even ignoring the leaf attack may be sufficient in most situations, chemical control of the pest is recommended during outbreak situation using Carbaryl 0.1%.

Coconut leaf roller : *Gangara thyrsis, Suastus gremius* : Hesperiidae

The skipper caterpillar is a pest of nursery plants and young palms in South India. *G.thyrsus* constructs rolls out of the leaflets usually from the tip, whereas *S. gremius* cuts one half of the leaflet and rolls into a case. Caterpillars feed on the leaf blades remaining within the leaf rolls leaving behind only the midrib. Caterpillar of *G.thyrsus* is stout bodied, pale green with reddish brown markings, but the body is covered with white waxy filaments. *S. gremius* caterpillar is green without any covering. Pupation takes place in leaf roll. Infestation of the pest is noticed more during monsoon months. Adults of both the species are chocolate brown butterflies with yellow spots on the wings, the size being much smaller in the case of *S. gremius*. The pest can be easily managed by hand collection and destruction.

Bark weevil : *Diocalandra stigmaticollis* : Curculionidae

The weevil infests the cuts and wounds on the tree and the petioles. It is more commonly seen on palms with stem bleeding disease. It burrows into the soft tissues of the bark. Weevils also tunnel within the base of the leaf petiole which is close to the stem. As a result, the fibres of the petiole base get loose and petiole looses its grip to the stem and fronds hang from the tree. In affected palm, reddening of the petioles and trunk especially around the wounds is noticed; production of leaf and spathe is delayed. In severe cases 9-10 fronds are seen hanging from the tree and trees become weak and yield reduces. Large number of adult weevils is seen in damaged palms within the base of the fronds. Adult is a small brownish weevil about 6 mm long. This can be effectively managed by spraying or swabbing the tree trunks with 0.05% Quinalphos after removing the hanging fronds.

Shot hole borer : *Xyleborus perforans* : Scolytidae

The small blackish beetles feed on the coconut stem making numerous minute holes on the bark. Presence of lot of minute holes interconnected by fibrous galleries on the tree trunk is the typical symptom of attack. Infestation is more during summer months. Swabbing the bark with 0.1% chlorpyriphos is effective in managing the pest.

Lace wing bug : *Stephanitis typicus* : Tingidae

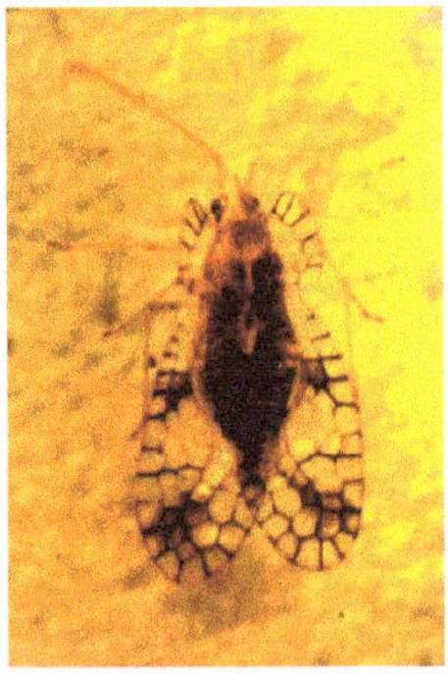

The nymph and adult bugs are gregarious in their habit. They feed by sucking the sap from the undersurface of the leaflets causing white specks on the upper surface which later turns yellowish. The insect is considered as a vector of mycoplasma causing the dreaded root wilt disease of coconut in Kerala. The population of the lacewing bug is highest during March to May and lowest in December. Adults are with transparent wings and netted venation. Nymphs are whitish with dark patches. The pest has a wide range of host plants which includes banana, ginger, turmeric, cardamom etc. *Stethoconus praefectus*, a mirid bug is the major predator of the nymph and adult bugs.

Nut borers : *Cyclodes omma* : Noctuidae

The caterpillars of these moths bore into the developing buttons through the perianth and feed on inner tissues causing nut fall.

Termite : *Odontotermes obesus* : Termitidae

Nymph and adult termite damages seedlings in the nursery and transplanted seedlings. They feed on the collar region of the seedlings and result in wilting of the central shoot. In grown up palms the termites attack the bark and runways made of soil and fibre are seen plastered on the tree trunk. The damage is more severe in the West Coast region .Drenching the nursery with 0.05% chlorpyriphos twice at 20-25 days interval manages the pest in the nursery. Infestation on older palms can be managed by spraying 0.05% chlorpyriphos on the affected trunk. Normally, the pest attack is best ignored and sometimes, a water spray on the trunk is given while irrigating the palms

Leaf Mites : *Raoiella indica, Oligonychus iseilemae, Tetranychus ludeni*

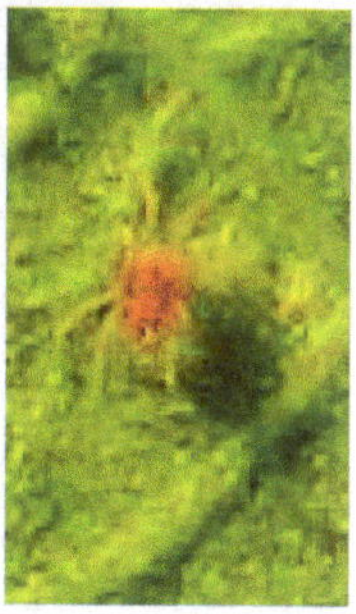

The foliage mites colonize the lower surface of leaves; suck sap leading to drying of the affected portion. Seedlings are more prone to the infes-tation. Mite infestation peaks during summer months. Spray application of 0.05% dicofol or dimethoate controls the mite effectively in seedlings. During summer months provide adequate shade and irrigation to reduce the mite incidence in seedlings.

The orange mite infests perianth region on nuts causing severe nut fall.

Rats : *Rattus rattus wroughtoni; Bandicota bengalensis* : Muridae

The rats cause extensive damage to tender nuts. Living on the crown of palms they make entry hole in developing nuts and feed on internal contents and live inside burrows in the coconut plantations. Management is by keeping single dose anticoagulants like the bromadiolone cakes on the crown and also on the ground, continuously until the time when the rat cake is no longer eaten. When the trees are not touching each other, providing a band with cellotape will prevent the rat from climbing the tree.

IPM in Coconut

IPM in coconut stresses the need for clean cultivation. The premises should not contain any decaying debris or cow dung or cut and decaying stem etc. These help to harbour grubs of rhinoceros beetle, pupae of black headed caterpillar and other life stages of pests. Compost pits and manure pits should be surface treated with dusts of carbaryl or lindane @ 0.01% v/v or the green muscardine fungus. This should be repeated once in three months. After every 3 months, the innermost leaf axils of trees have to be filled with a mixture of lindane/ carbaryl dust with sand to ward off attack by rhinoceros beetle and coreid bug. Cut holes on the stem should be sealed with mud slurry containing carbaryl, to prevent egg laying by red palm weevil.

Against eriophyid mite attacking the buttons and attack by other mites, apply Dicofol @5 ml/l or Dimethoate 1.5 ml/l or Neem oil/garlic extract/soap suspension (20:20:5g for 1 litre)once in three months on the first three bunches after pollination. Hook out adult rhinoceros beetles immediately on spotting the attack. Dried and old leaves in the garden should be removed as firewood or burnt, to destroy pupae of leaf eating caterpillar. Coconut log traps can be set up longitudinally splitting cut stumps (50cm long) and pouring toddy or fermented coconut water into the central pith. The adults of rhinoceros beetle or red palm weevil collecting on them should be destroyed every day. Apply 2 gm carbaryl solution in one litre to the pith once a fortnight to kill the grubs of scarabeids /weevils developing on the pith. Set up pheromone traps 50m away from the garden in non-coconut area or at least in the outermost periphery of the garden. Collect and kill the adults everyday, provide pineapple waste or fermented coconut water or sugarcane splits in the trap, with 1ml Malathion. This bait should be removed and fresh bait provided every week.

Release parasitoids like *Bracon brevicornis* @ 1800/acre or *Goniozus (Parasierola) nephantidis* @1200/acre once every month during summer months when the pest attack is heavy.

Root grubs can be managed by applying chlorpyriphos @ 2.5 ml/l to the root zone just after the pre-monsoon showers. Set up light traps during dusk in the periphery of the garden during pre-monsoon showers when the adults emerge. Burn trash, old dried leaves, petioles and leaf bases during dusk to attract and kill adults.

Chapter - 51

COFFEE

Coffee is an important cash and export crop for small-scale farmers. The crop suffers heavy yield losses due to damage caused by a wide range of insect pests

White stem-borer : *Xylotrechus quadripes* : Cerambycidae

This is one of the important and serious pests of *arabica* coffee. Other types are also attacked to a slightly lesser extent. The adult is a dark-brown longicorn beetle having several white markings on the elytra and thorax.

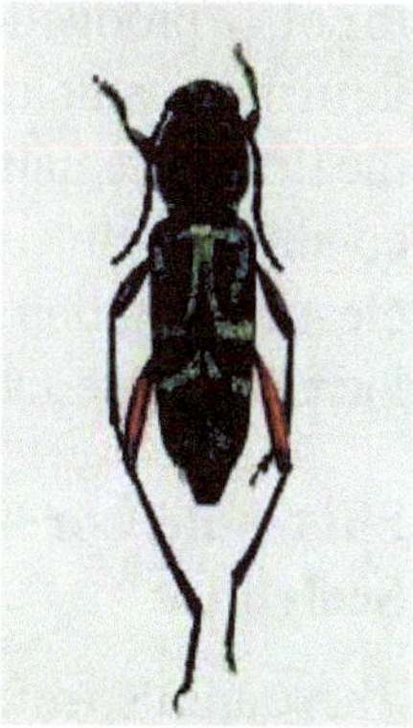

The adults start emerging in April-May and lay eggs during October and November in small groups on the stem under the bark or under the moss growing on the stem. The eggs hatch in 9-15 days. The emerging grubs bore into the bark and tunnel in all directions and feed on the internal tissues. The grub is pale yellow, stout and with a swollen head. Pupation is within the tunnel and comes out after around a month, through an exit hole that was made by the grub. The beetle has an annual life cycle.

The stems and branches of the plants are killed as a result of the grubs tunneling within them. Young bushes suffer the most and the older plants lose their vigour and yielding capacity. Infested plants show external ridges around the stem. Affected plants also show yellowing and wilting of leaves.

Management

Uproot and destroy the severely infested plants prior to the emergence of the adults during March to May. Loose scaly bark on main stem and thick primaries may be scraped and removed.

In endemic situations, spray an insecticide like chlorpyriphos, quinalphos or Malathion on the stem at 0.05% concentration during the egg laying period to prevent oviposition by the adults.

In case of heavy incidence, install pheromone traps @ 25 traps/Ha.

Coffee berry-borer : *Hypothenemus hampei* : Scolytidae

Pin holes at the tip of berries indicate presence of the pest. In severe cases of infestation two or more holes may be seen. Infested berries may fall due to injury or secondary infection. Severe infestation may result in heavy crop loss.

The coffee bean borers, *Hypothenemus hampei/ Step-hanoderes coffeae* pose a significant threat to coffee production in many estates. Poor care and lack of sanitation on the farm may result in cherries left on the plant to mature or dry, or those fallen to the floor to remain undisturbed and remain till the following season. These conditions attract the beetles, breeding in the dry, more compact tissue. The clearing of fallen cherries and removal of older specimens can prevent the increase of infestation.

Shot hole borer : *Xyleborus morstafti*; *Xylosandrus compactus* : Scolytidae

These are also referred to as ambrosia beetles. These shot-hole borers are sometimes serious pests of the robusta coffee. The female beetles infest underside of the succulent tertiary branches at the internodes. The beetles make pin holes on the twigs leading to tunnels within. Once inside they excavate galleries and lay eggs.

This excavation, along with the introduction of pathogens, is the cause of damage to the host. Damage is not caused by feeding, since the beetle larvae feed on *ambrosia* fungus introduced by the female. Infestation by one to three females is sufficient to kill the twig or branch. The attacked twigs show discolouration around the pinholes, drooping leaves which turn yellow and fall off and 'die-back' symptoms appear. The terminal leaves wilt, droop and dry up. Opening of the side branches are covered with a wad of damp sawdust, which helps in the growth of *Ambrosia* fungus, *(Monacrosporium ambrosium)* inside the galleries, on which the larvae feed. Spores of this fungus are carried by the female beetles in special cavities at the base of the mandibles called mycangia. The galleries are lined by this fungus. Male larvae are much smaller than the female larvae. The beetles are vectors of *Fusarium* sp. The pest infestation is heavy in November and December after the monsoon.

Management

- Pruning and destruction of beetle-infested plant material is essential. Good care to promote plant vigor and health will help in resisting infestation or recovering from infestation.
- Mass-trapping of adult beetles is an option currently being investigated, using various types of traps and attractants
- Prune affected twigs 5 cm- 7.5 cm below the hole
- Remove and destroy all unwanted/infested parts during summer
- Maintain thin shade

Spotted grasshopper/Coffee locust : *Aularches miliaris* : Acrididae

This is an occasional pest assuming locust proportions in some years, feeding on shade trees, areca nut, banana and wild plants. Nymphs and adults feed gregariously on the vegetation including cardamom leaves, and produce a frothy secretion between the head and thorax when challenged. Adults are green and beautiful to look at, and spotted with yellow markings.

In the swarm phase, rake the egg pods and expose the eggs in the soil, collect and destroy the nymphs and adults, and if needed, spray quinalphos 0.05 percent or dichlorvos 0.75 per cent.

Mealybug : *Planococcus citri* and *P. lilacinus* : Pseudococcidae

Mealy bugs are the major pest of *arabica* coffee. It infests the roots and some parts of the shoots leading to debilitation of the plant and crop loss. Spongy tissues develop on the roots infested by the bug. On the shoot it infests the underside of leaves, the base of leaf petioles, spikes, nodes, fruits and the tender branches. The leaves become chlorotic and fall, spikes are suppressed and the fruits drop. Ant activity is noticeable on the plants affected by mealy bugs.

Management

- Control ants by dusting quinalphos (1.5%) or methyl parathion (2%) or malathion (5%) in the base of shade trees
- Destroy ant nests, remove and destroy weeds
- Spray affected patches with quinalphos 25EC or Fenitrothion 50EC @ 300 ml each or fenthion 1000 @150 ml .Add liquid soap @4-5 ml/l for enhanced effectiveness of the spray.
- Release parasitoid *Leptomastix dactylopii* against *P. citri* or predator *Cryptolaemus montrouzieri*

Green Scale : *Coccus viridis=(Lecanium viride)* : Coccidae

This insect is also referred to as the coffee green scale. Attack by the scale leads to debilitation of older plants and death of nursery plants. Honey dew secreted by the scale forms a film on leaves and a black fungus (sooty mould) grows on it. Ant activity noticed in the affected twigs and branches.

The adult female is shiny pale green with a conspicuous black, irregular U-shaped internal marking that is dorsally visible to the naked eye. Two sub-marginal black eye spots are also present and can be seen with a hand lens. Dead scales are light brown or buff color and the black internal marking is lost. Nymphs, or immature green scales are oval, flat and yellowish green in color, and have six short legs. There are three nymphal stages before becoming an adult.

Management

Several entomogenous fungi were observed associated with green scale on citrus, and some apparently played an important role in the natural limitations of the scale on citrus during certain seasons of the year. These fungi include: the white-fringed fungus, *Verticillium* (*Cephalosporium*) *lecanii* (Zimmerman); *Empusa lecani* Zimm.; *Aschersonia cubensis*; the pink scale fungus, *Nectria diploa*; and a grayish blue fungus, The white-fringed fungus is the most common and apparently causes the highest percentage of mortality. In the field, there are also many parasitoids and predators for the pest.

In management, maintain optimum shade. insecticidal sprays, containing 5ml/ l (0.5%) liquid soap should be directed to the lower leaf surfaces and new growth to give thorough coverage. This should also be directed to the ant nests. Insecticides that can be used include quinalphos (1.5 %D) or methyl parathion (2%D) or malathion (5%D)or other spray formulations like quinalphos 25 EC or fenthion or dimethoate 30EC or fenitrothion 50 EC with 200 litres of water and400-500 ml of soap solution.

Brown scale : *Saissetia coffeae* : Coccidae

Damage similar to that caused by mealy bugs and green scale.

Cockchafers or white grubs : *Holotrichia* spp. : Melolonthidae

Yellowing of leaves and stunted growth. Affected plants can be easily pulled out and may wilt and dry.

Cockchafer grubs can be killed by drenching with 0.1% lindane or chlorpyriphos, one to two months after pre monsoon showers where first instar grubs are seen as surface soil root feeders

Coffee Bean borer : *Areaecerus fasciculatus*: Anthribidae

Though the pest attacks berries in the field, severe infestation occurs only in storage. Infested bean show a circular hole. Attacked berries in field shrink and become black. The adults and larvae make feeding tunnels in all directions in the beans. The stores can be fumigated with aluminium

phosphide. Most important storage precaution is to dry the beans below 8% moisture.

Hairy caterpillars : *Eupteroet canarica and E. fabia;* Eupterotidae.

Caterpillars feed on leaves and cause severe damage by denuding the plants.

Red borer : *Zeuzera coffeae* : Cossidae

In the early stages of attack, young plants/branches show sign of wilting. The infested plants usually bear one or two holes through which pellet like excrement of the larva hangs out, and accumulate at the base of the plant. In advanced cases, branch of the whole plant dries up. Adult of orange-red borer is a medium sized moth with many black spots and patches on both the wings, but more on the forewings. Eggs are laid in strings on the barks. Incubation period is 8 to 12 days. Larva is reddish, and hence the name red borer. After entering through the junction of leaf stalk and twig, larva constructs a tunnel in young plants and starts boring itself down to the root region. Larval and pupal stage together last for about 12 to 24 months. After moth emergence, pupal skin protrudes outside through exit hole.

Management

- Cut and burn red borer infested plant or twigs
- Systematically monitor the plants for the presence of the borer, especially the pellet like excreta in the soil or on the plant. Spray the bark or trunk with a contact insecticide in and around the infested plants and in endemic attacks.
- The entomopathogenic fungus *B. bassiana* can be used against the pest on a continued basis .

Snail : *Ariophanta solata*

Feed on leaves of Arabica coffee and dadap, and occasionally on the bark of tender branches and the skin of fruits.

Thrips : *Heliothrips haemorrhoidalis; Haemorrhoidalis; Retithrips syriacus; Scirtopthrips bispinosus; Thrips nilgiriensis*

Tiny fragile insects that lacerate and feed on leaves; affected leaves become distorted, crinkled and mottled; severe in hot weather, though found throughout the year Lacerations appear as brownish irregular streaks or silvery patches on the leaf surface. The Damage is more pronounced in exposed areas during dry weather.

Remove weeds and grasses that harbour thrips; spray with 0.03% Dimethoate, Methyl demeton or Diazinon

Mites : *Oligonychus coffeae*; *Brevipalpus obovatus*; *B.phoenicis*

Mites feed in large colonies on the under-surface of the leaves; attacked leaves dry up in patches

Management

Dust sulphur or spray wettable sulphur (0.05%)

Root-lesion nematode : *Pratylenchus coffeae*

Affected young plants are lean and lanky. Older leaves become yellow and drop, leaving very few undersized, chlorotic and crinkled leaves at the tip of the stem giving a 'tufted' appearance. Affected bearing plants show thinner stem and have adequate foliage to support the crop. Affected plants are left with two or three pairs of leaves at the tip producing a 'goose-neck' condition after the receipt of rains.

- In nursery dig and expose the soil for one summer and thoroughly dry the jungle soil while preparing nursery mixture
- In the main filed, uproot and burn the affected plants, dig the affected soil and expose for one year
- Take care to keep the pits free from weeds- Plant the affected area grafted plants of *arabica* on *robusta* rootstock.

Berry borer : *Homona coffearia* : Tortricidae

The larvae feed on the tender leaves and bores into the tender berries.

Leaf miner : *Acrocercops caerulea* : Gracillaridae

The pale orange yellow larvae of this moth mine the coffee leaves on the upper surface in serpentine mines and feed on the chlorophyll. The leaves are shed prematurely.

IPM in Coffee

Berry borer, cockchafers, mealy bugs and scale insects are the main pests in coffee.

Coffee berry borer management requires complete avoidance of left over berries that would allow residual population to develop. Hence, pick the main crop completely and remove and destroy all unmarketable berries. This is a must in coffee IPM. Avoid exess shade, dry the berries to 8-10% moisture, and set up alcohol traps (Ethyl: Methyl alcohol 1:1) to attract adults and kill them. In heavy incidence areas, spray endosulfan 35 EC @ 350ml/acre at 120- 150 days after flowering. Where endosulfan is banned, malathion or quinalphos can be used. This would also kill leaf and flower feeding caterpillars, cockchafer adults, and other surface feeders. Add sandovit or other wetting agent @ 300ml-400ml. This would ensure management of scale insects and mealy bugs. Specific chemicals for mealy bugs include Quinalphos 300ml, fenitrothion 300ml or fenthion 150ml per 200 l spray fluid. Mealy bug and scale management requires ant control using dusts of quinalphos or malathion or methyl parathion, which also kills surface feeders.Release the ladybird beetle *Cryptolaemus montrouzieri* against mealy bugs and scales and also spray *Verticillium lecani* @ 6x106 spores/ml. Cockchafer grubs can be killed by drenching with lindane 20 EC @ 750 ml in 200 l of water per acre, one to two months after pre monsoon showers where first instar grubs are seen as surface soil root feeders.

Chapter - 52

OILPALM

Spindle bug: *Carvalhoia arecae* :Miridae

Nymphs and adult bugs suck the plant sap from spindle leaf and tender leaves. Feeding marks are seen as necrotic lesions in the spindle and fully opened leaves. The pest is seen in the nursery plants round the year but incidence is high during rainy season.

Tussock caterpillar: *Dasychira mendosa*: Lymantridae

Caterpillars of this polyphagous pest feed on young and mature leaves of seedlings, causing considerable defoliation. The incidence of the pest is more during June and July after a few heavy rains.

Rhinoceros beetle: *Oryctes rhinoceros*: Scarabidae

The adult beetle bores through into the spear leaf, resulting in snapping of the fronds at the feeding sites. The beetles also bore into the male and female inflorescence even when they are within the spathe.

Red palm weevil :*Rhynchophorus ferrugineus* :Curculionidae

The adults of the beetle are seen feeding on the meristem inside the stem and also causing tunneling of the stem, toppling the trees, especially in young trees.

Termite: *Odontotermes* sp. :Termitidae

The termites attack the roots of seedlings, and also occasionally, the inflorescence spear and fruit bunches.

White grub: *Apogonia* sp., *Adoretusa* sp.: Melolonthidae

The grubs of these beetles have an annual life cycle. They emerge just after the monsoon, lay the eggs in soil, hatch out and feed on the roots of oil palm and other feeder roots of all plants, completing the larval life in around nine months, and then pupate inside the soil. The seedlings may require an application of insecticide dusting to protect their roots.

Leaf caterpillar: *Spodoptera litura* :Noctuidae

They are defoliators on the seedlings and occasionally on the larger fronds also.

Scales: *Pinnapsis aspidistrae, Hemiberlesia latariae, Chrysomphalus aonidum*: Diaspididae

The scales are seen feeding on the bunches of oil palm occasionally.

***Meal bugs:Dysmicoccus brevipus, Palmiculator* sp.,: Pseudococcidae**

The mealy bugs also are seen congregating on the male and female inflorescences and also on the ripening bunches, desaping them and causing reduction in fruit quality.

Bagworm: *Manatha albipes, Metisa plana* :Psychidae

The caterpillars of these species of bagworms cause severe defoliation occasionally on oil palm.

Vertebrate pests: *Crows, parrots* and *babbler*

They feed on ripe fruits and destroy them.

Rats: *Bandicota indica,Rattus rattus wroughtoni* : Muridae

The rats tunnel into the bole of the seedlings causing their death. They also damage ripe bunches and spikelet of male inflorescence.

Chapter - 53

TEA

Tea tortrix/flushworm: *Homona coffearia*: Tortricidae

The tortrix is an important pest of tea leaves. The larvae make leaf nests by webbing the leaves using silken threads and feed from inside. A single caterpillar makes several cases. Young larvae prefer tender leaves, while the older larvae are seen in mature leaves. The economic produce being the leaves, and since chemical control is not advised, especially during the picking phase, the pest becomes a management problem.

Biology

Adult moth is bell shaped in outline while at rest, and is light brownish, around 3 cm in wing span. Forewings have two wavy transverse bands, one towards the middle and another on the apical end. Patterns of female are less distinct and have wavy margins instead. Eggs are laid in masses of 100-150 in clusters on the upper surface of mature leaves. Male is smaller. Egg period is 6 – 8 days. The larvae fold the leaf by webbing the margins by silken threads and feeds from the inside. Larva is green in colour with a black head and larval period is 3-4 weeks.. Pupation takes place inside the leaf cases it self. Pupa is green initially and turns to reddish brown. Pupal period ranges from 9 – 15 days. The nursery plants are most attacked, where it is managed by sprays of endosulfan, phosalone, chlorpyriphos, dimethoate, fenitrothion, malathion or phenthoate, @ about 200-300 ml/ha.

Tea leaf roller: *Caloptilia (Gracillaria) theivora*: Gracillariidae

The moth is tiny, purple brown, and sits in a mosquito posture with its lean wings folded back. They are micro lepidoptera and have the anten-na longer than body, with golden iridescent patches in forewing and abdomen. The larvae that hatch out remains as a leaf miner up to the second instar and later they larva roll the leaves from tip downwards. Pupation is in a silken cocoon on the underside of the leaf. Management measures for flush worms manage this pest also. Endosulfan, phosalone, chlorpyriphos, dimethoate, fenitrothion, malathion or phenthoate can suppress the pest. Usually, removing and destroying the affected leaves with the larvae inside, itself can manage the pest.

Flushworm/leaf folder :*Cydia (Laspeyresia)leucostoma* : Tortricidae

The larvae of this tiny moth are called as the flush worm since it starts feeding on the new flushes that grow after clipping. The moth is blackish brown in colour, with yellow and white streaks on the forewings. Eggs are pale yellow and laid singly on the under surface of mature leaves. Incubation period is 4-5 days. Larva is brown and larval period is 19-25 days. Caterpillar webs together, the margin of tender leaves and forms a case enclosing the bud. Affected leaves be-come rough, discoloured and leathery. If the bud is eaten, the shoot growth is stopped. Pupation takes place on a fold made on the older leaves. Pupal period is 8 – 10 days.

Management is done by periodic collection and destruction of the webbed bud and leaves. Endosulfan, phosalone, chlorpyriphos, dimethoate, fenitrothion, malathion or phenthoate 300-400 ml/ ha may be applied in severe attack.

Leaf webber :*Ereboenis saturata*: Pyralidae

The caterpillars are leaf webbers.

Bagworm: *Clania crameri; Pteroma* sp. F: Psychidae

The larvae of this pest feed on the surface tissues of the leaves. The males are winged and females are worm like.

Female of bagworm ***Clania crameri* See the head protruding out of the case**

Slug caterpillar: *Latoia* (*Parasa) lepida*) :Cochlididae

These feed on the leaves occasionally.

Tea mosquito bug: *Helopeltis antonii, H.theivora*:**Miridae:**

It is a serious pest of tea and widely distributed in southern India. The alternate hosts include Cashew, neem, moringa and guava .

Adult is black in colour with a mixed red and black colouration. Body is small, slender with long antennae. An erect pin like knobbed process on the scutellum is characteristic of the species.

Bugs are active in the early morning and late evening hours and hide amongst the bushes during the remaining period. Eggs are inserted into tender shoot, and are provided with two filaments arising from the operculum. Egg period is 5-7 days. Nymphs are red with long legs and antennae and resemble ants. They have five nymphal instars. Nymphal period is 15 to 20 days. Adults and nymphs suck the sap from buds, young leaves and tender stems by puncturing the plant tissues with needle like stylets and inject toxic saliva. These punctures appear as reddish brown spots. Due to intensive feeding,spots develop on the leaves and later they curl up, become badly deformed and remain small. Gradually, shoots dry up.

Management.

1. Collect and destroy the damaged plant parts
2. Spray endosulfan or phosalone or chlorpyriphos or dimethoate or

fenitrothion at 2ml / lit of water. Spray in early morning or late evening hours on trunks, branches, foliage and inflorescence to manage the active nymphs and adults.

Aphid : *Toxoptera aurantii* : Aphididae

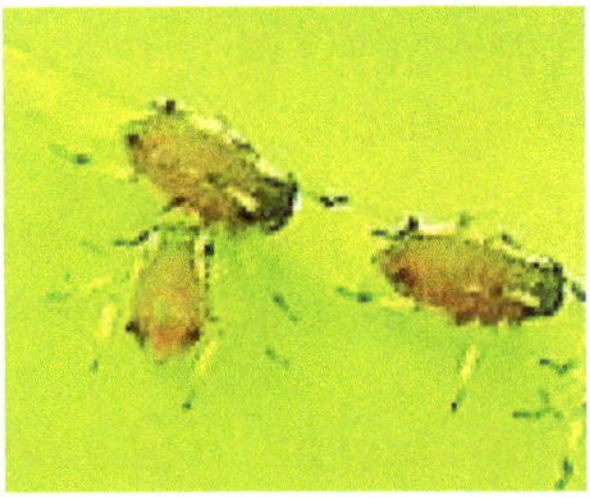

It is a polyphagous species attacking tea and other host plants such as coffee, cacao, citrus etc. Colonies of this dark brown or black aphid are seen on tender shoots, leaves and buds of tea immediately after pruning. Due to feeding by aphids, leaves curl up and shoot growth is stunted. Ants attend on aphids. Honey dew excreted by the aphids falls on the leaves and facilitates the growth of sooty mould fungus. Adult is dark brown in colour. Both alate and apterous forms exist. The seedlings and young plants are more affected.

Scales: :*Saissetia coffeae* : Coccidae

These hemispherical brown scales are seen along the midrib and tender stem. Sooty mould is seen on lower leaves after the attack. The vegetatively propagated clones are more susceptible.

Nymph is white; Adult male is winged and the female is sedentary. If the attack is severe, spray carbaryl 2 g/lit, endosulfan 2 ml/lit, quinalphos 2 ml/lit and if unmanaged, use chlorpyriphos 2 ml/lit.

Sapling borer: *Sahyadrassus malabaricus*: Hepialidae

Presence of the pest is understood when chewed tissue is seen at at the col-lar region. The mouth of the tunnel in which the borer feeds show pieces of bark, wood and frass particles held together by silk. The saplings break off at the point of injury.

Adult moth is peculiar and hangs vertically down by the support of two pairs of legs. Egg period is for 7-10 days, and larvae live for about 10 months inside the larval tunnels. They pupate for 3-5 weeks.

Management is done by monitoring, and cleaning the base of bush. The hiding larvae inside the tunnels are mechanically killed. Injection of insecticides like quinalphos @ 2ml/l and plugging the holes can also be practiced.

Shot hole borer/Ambrosia beetles: *Euvallacea (Xyleborus) fornicates* : Scolytidae

Adult female is black in colour, small about 20-25 mm only. Male is half the size of female, generally described as devoid of wings, but more appropriately as flightless. Males are very rare, the sex ratio being around 1:8. The egg period is 4 – 6 days and there are three larval instars that last for 16 – 18 days. Pupae are whitish and pupal period is 7 – 9 days. Total life cycle is 27 – 33 days. The females bore into the sapwood making vertical tunnels with side branches within it. Opening of the side branches are covered with a wad of damp sawdust, which helps in the growth of *Ambrosia* fungus, *(Monacrosporium ambrosium)* inside the galleries, on which the larvae feed. Spores of this fungus are carried by the female beetles in special cavities at the base of the mandibles called mycangia. The galleries are lined by this fungus. Male larvae are much smaller than the female larvae.

Attack by the beetles is seen as presence of round shot holes in primary branches. The branches show dieback and the buds die. Circular or longitudinal tunnels present inside the stem. New branches in the first year of pruning are devoid of attack. Pest builds up by second year and is more pronounced in the third year and further. The attack debilitates the crop, causing galleries even in the collar region, and poor bush frames, if kept uncared for.

Spray Chlorpyriphos 20 EC @0.05% directed towards the base of the plant towards the end of the first year and repeat the sprays, once every three months, if need be.

Thrips: *Scirtothips bispinosus*:Thripidae

Thrips lacerate leaf tissues and suck in oozing sap. Lacerations appear as brownish irregular streaks or silvery patches on the leaf surface. Infested leaves become distorted, crinkled or mottled. Damage is more pronounced in exposed areas during dry weather.

Adults are minute, with fringed wings . Thrips insert eggs singly into the leaf tissues. Eggs hatch in 9 days. There are two nymphal instars followed by pre pupal and pupal stages. Thrips are summer pests and get reduced by maintaining optimum shade. If the attack is severe spray Imidachloprid20 EC2.5-3ml/10 L, Thiacloprid 0.5-one ml/L, Fipronyl 5% 2 ml/L or Endosulfan35EC 3ml/l. Provide sprinkler irrigation or irrigate the whole plants just as a rain.

Red borer: *Zeuzera cof-feae*: Cossidae

Red borer has a wide range of hosts including coffee,tea, cinnamon, sandal, cotton, orange, teak and many forest trees.

Larva causes damage by boring the stem or branches to feed on the wood. In early stages of attack, young plants or branches show signs of wilting. Infested part bears one or two holes through which, pellet-like excrement of the larva hangs out and accumulate at the base of the plant. In advanced cases, the branch or the whole plant dries up.

Adult of orange-red borer is a medium sized moth with many black spots and patches on both the wings, but more on the forewings. Eggs are laid in strings on the barks. Incubation period is 8 to 12 days. Larva is reddish, and hence the name red borer. After entering through the junction of leaf stalk and twig, larva constructs a tunnel in young plants and starts boring itself down to the root region. Larval and pupal stage together last for about 12 to 24 months. After moth emergence, pupal skin protrudes outside through exit hole.

Management

1. Cut and burn red borer infested plant or twigs
2. Systematically monitor the plants for the presence of the borer, especially the pellet like excreta in the soil or on the plant. Spray the bark or trunk with a contact insecticide in and around the infested plants and in endemic attacks.
3. The entomopathogenic fungus *B. bassiana* can be used against the pest on a continued basis.

Bark borer: *Indarbela theivora*: Arbeliidae

Eggs are laid in cuts or cracks in bark of trunk or main branches of tree. On hatching, caterpillars feed superficially below bark, making zigzag galleries and later bore inside bark, or main stem. They remain within these burrows during day and come out at night and feed on bark. Symptom of attack is presence of huge silken webbed masses comprising chewed wooden particles and excreta, which hang loosely on tree trunk covering the entry holes. Pupation takes place inside the tunnels. There is only one caterpillar or pupa in each burrow, but there may be several such holes and burrows, on a single tree.

Full-grown caterpillars are stout and dirty brown in colour. Pupae are also stout, reddish-brown in colour. Moths are pale brown. Hind wings are whitish. Egg period is 6 to 8 days, while caterpillar stage is extremely long and takes 10 to 11 months. Pupal period is completed in 14 to 20 days. There is only one generation in a year.

Caterpillars of these are bark borers that make galleries of silk and frass on the bark and feed within.

Looper caterpillar: *Buzura (Biston) suppressaria*: Geometridae

Adult is a straw coloured moth. Wings are grey with light yellowish brown markings and wavy lines along the outer margins. The grey or dark green larvae are defoliators on tea.

Lobster caterpillar/crab caterpillar: *Neostauropus (Stauropus) alternus*: Notodontidae

The larvae are peculiar, with elongated thoracic legs along with elongate abdominal claspers, together which makes it resemble crab legs and hence the name. They also feed on the leaves.

Tea Mites

Red spider mite: *Oligonychus coffeae*: Tetranychidae

It is an important mite species in tea plantations in India. Nymphs and adults infest upper surface of older leaves. Infestation starts inside the shelter of webbings along midrib and veins and gradually spreads to the entire upper surface of leaves. Affected leaves become bronzed, dry and crumpled, and then fall off. They are summer pests and get washed away in rains.

Adult female is elliptical in shape, bright crimson red anteriorly and dark purplish brown posteriorly. *O. coffeae* is the largest of all tea mites. These mites spin a web of silken threads on the older leaf and protect themselves. Eggs are reddish, spherical and are laid singly along midrib and veins. Egg and larval periods last for 4 – 6 and 6 – 8 days respectively. Life cycle from egg to adult is 10 – 14 days. In severe cases, young leaves may also be attacked. These are managed by the application of sulphur dust(80 WP, 5gm/l) or Kelthane @ 1.5-3ml, based on intensity of attack.

Scarlet mite : *Brevipalpus californicus (=australis), B. phoenecis* : Tenuipalpidae

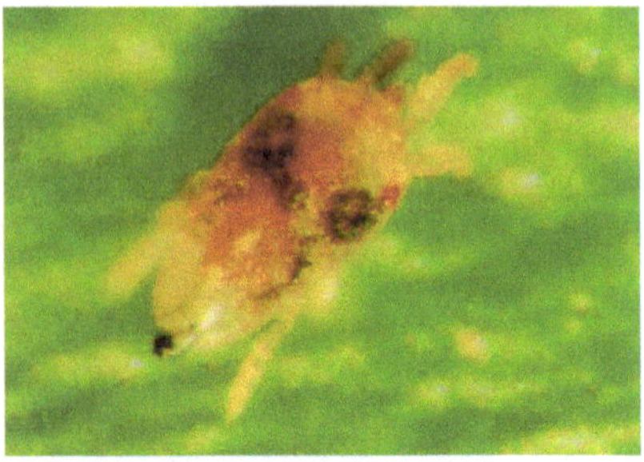

Adult mite is flat, elongated, oval and scarlet red in colour with dark marks on the dorsum. They can be seen with the naked eye. Eggs are bright red, elliptical and laid in clusters on the undersurface of leaves. Around 45-50 eggs are laid. Nymphs enter a quiescent stage before becoming the adults. The total life cycle lasts about one month. They congregate on the under surface of mature tea leaves especially along the midrib and petiole region. Feeding by scarlet mites leads to brown discoloration of leaves, splitting of leaves and drying.

Pink mite or orange mite: *Acaphylla theae*: Eriophyidae

Acari

It is an important mite species in India. These microscopic mites are found on both sides of the young leaves, petioles and tender stems. Affected leaves turn pale and curl upward. In severe infestation, leaves become discoloured, leathery and brown. The veins show pink discolouration. The bushes show a sick appearance. Damage is restricted to top 10 – 15 cm of tender leaves. Assam type of tea is more susceptible.

Purple mite: *Calacarus carinatus*: Eriophyidae

This eriophyid is a major pest in south India tea plantations. It is prevalent on the both surfaces of mature leaves. The surface of leaves damaged by purple mites is characterized by the coppery brown discoloration.

Adult mites are very small and spindle shaped dark purple in colour with five longitudinal white waxy ridges on the dorsal side. Eggs are sub-hemispherical, transparent and are laid singly on the under surface of leaves. Egg period is 3 – 5 days and nymphal period is 3 – 6 days.

Yellow mite: *Polyphagotarsonemus latus* : Tarasonemidae

They are polyphagous species attacking tomato, cotton and pulses. They are also seen on young tea leaves especially the top two to three leaves and the bud. Affected leaves become rough, brittle and corky lines or patches appear on the lamina. Internodes get shortened, shoots become stunted and deformed.

Mites are pale yellow in colour, with the male shorter than female. Fourth pair of legs are seen and are provided with a curved tooth and a pair of whips.

IPM for mites

1. Collect and destroy all types of damaged parts along with mites

2. Spray ethion 50 EC 2 ml/l or dicofol 18.5 EC 2 ml/l or wettable sulphur 80 WP 4 g/l or phosalone 2 ml/l or thiometon 1.5 ml/l using hand operated sprayer to ensure proper coverage of spray solution on both sides of the leaves.

Chapter - 54

BETELVINE

Scale insect: *Lepidosaphes* sp.: Diaspididae

The mature scale is elongated and oval in shape and dark grey in colour. They infest the stems, leaves and petioles, causing shriveled leaves due to sucking of the sap. The affected parts wither and dry up. In severe infestation leaves dry up and give sticky look and gradually the vine wilts.

Betelvine bug: *Disphinctus politus*

It is a serious pest in many parts of the betelvine areas. Tender leaves are punctured by the insect while feeding on the leaf and this results in a number of minute black spots all over the leaves. When such leaves mature, these spots result in small holes. Very badly affected leaves curl, shrivel and give a puckery appearance. The bug is active during June- October and multiples rapidly under warm and humid conditions. They take shelter in dark and protected places and thrive by feeding on the plant sap.

Mealy bug: *Ferrisia virgata*: Pseudococcidae

White cottony mealy bug causes considerable damage to the tender leaf petioles and apical shoots mainly during rainy season. Nymphs and adults suck the sap from leaves and twigs. Desapping causes white patches on the leaves which are visible from a distance.

White fly: *Dialeurodes pallida* : Aleyrodidae

This is a serious pest in the eastern and north eastern regions of India. Both adults and nymphs of the insect suck the sap from lower surface of leaves resulting in pronounced chlorosis. Sooty mould develops on honey dew excreted by the insects.

Red spider mite: *Tetranychus* sp.Tetranychidae

All the stages of the pest from nymphs to adult feed on the under-surface of the leaves by sucking the leaf sap. Characteristic blotching of leaves develop due to removal of chlorophyll and other plant pigments. The leaves present a pale yellow look which dries off after some time.

Chapter - 55

CARDAMOM

Cardamom thrips: ***Sciothrips cardamomi*****: Thripidae**

Thysanoptera

This thrips attacks other crops like amomum, tea, colocasia, grapevine, castor, cotton, ginger and turmeric also. The thrips are minute greyish brown insects with fringed wings. They breed on unopened leaves, spindles, leaf sheath, flower bracts and floral tubes .

Thrips lacerate the ovary and surface tissues of capsules and suck the exuding sap. The injured tissues form a corky layer on the capsule surface which appears as scabs. Such capsules appear stunted, malformed and shrivelled with encrustations on the skin. The condition is popularly known as "Cardamom Itch". The pest adversely affects the shape, appearance and aroma of the

capsules. Seeds from infested capsules are not pulpy, juicy black, and have poor germination. The attack also causes stunting of panicles and shedding of flowers. Scabbing of capsules reduce the quality and quantity up to 80-90%.

Adult females thrust the minute, kidney shaped eggs on the leaf sheath, flowers and surface tissues of capsules. Eggs hatch in 9-12 days. The nymphs grow by lacerating the surface of the tissues and feeding on the exuding sap. Nymphal period lasts for 9-12 days. Pupation is for 3-5 days. High temperature and low humidity favours the growth of the insects.

Management of thrips is very important in cardamom crop. Trashing, which involves removal of dry leaves and leaf sheaths is to be a regular practice. Destroy alternate hosts, including amomum, *Alpinia curcuma* and colocasia from the field. Four rounds of spray with fenthion, profenophos or quinalphos @0.05%and 0.04% triazophos may be needed if the attack is heavy.For organic cardamom, apply bio rational insecticides like fish oil insecticidal soap, wherever available, along with 2.5% tobacco extract at the basal one third of the plant. This may be repeated when the panicle emerges at the base of the culm.

Shoot, panicle and capsule borer: *Conogethes (Dichocrocis) punctiferalis*:Pyraustidae

Lepidoptera

This pest also attacks Castor, turmeric, ginger, pomegranate, custard apple, peach, plum, cocoa, guava, mulberry and even teak. The larva bores into the central core of the pseudostems, young tillers, panicles and capsules resulting in the death of the central spindle causing characteristic "dead heart" symptom. Larvae bore in to the immature capsules feeding on seeds and rendering them empty. Oozing out of excreta and frass at the mouth of the bore hole is very conspicuous on stem / pods. Adult is a pale, yellowish moth with black spots on both the wings. Loss in yield can be up to 40 per cent.

Adult moth lays eggs singly/ groups on tender parts of plant. The egg period lasts 6 to 7 days and emerges out into brown larvae that are covered with minute hairs arising on warts. They feed for 15-18 days. Pupation takes place in loose silken cocoon in larval tunnel. Pupal period lasts for 7-10 days and the life cycle is completed in 30-35 days.

For management of the pest, it is necessary to have a judicious application of nitrogenous fertilizers. Remove all alternate hosts as much as possible from the vicinity of the crop. Infested clumps should be rogued and destroyed. Set up pheromone traps @ five traps per acre. Insecticides recommended include Phosalone 35 EC at 3 ml/l, endosulfan 3ml/l,fenthion1.25ml/l,quinalphos 4ml/l, and profenophos 2ml/l.

Root grub: *Basilepta fulvicorne*: Eumolpidae

This pest has assumed significance from the late nineties as a grub causing serious yellowing and wilting of cardamom plants causing yield loss to the tune of 30-70 per cent. The adult beetle is around 5mm only, shining metallic green to blue, and may be seen nibbling the leaves of cardamom, especially in unshaded gardens. However, grub damage is reported to be less in adequately shaded plantations. The adults lay 150- 400 eggs in the soil, and they hatch out into larvae that are around 2 mm in size. These C shaped grubs, are also called root grubs and feed on any small rootlet, including those of jack, mango, fig, and even weeds. In cardamom, they also feed on the rhizome, predisposing it to rot fungi. Grub stage can be from 45 to 120 days depending on the season. They attain the size of about 10 mm upon completion of larval life and pupate in the soil. One life cycle is completed in about three months and the second brood may extend to more than six months.

Collection and destruction of adult beetles can be done during the mass emergence using hand nets. Insecticides recommended are fenthion, quinalphos, and chlorpyriphos on the leaves during the mass emergence of the adults. Soil drenching of imidachloprid, 0.015% or chlorpyriphos 0.07% @ 5 l/plant is to be done to prevent damage to the roots.

Cardamom whitefly: *Singhiella (Kanagarajiella) cardamomi* : Aleyrodidae: Hemiptera

Nature of damage

This is a pest that has become serious due to the indiscriminate use of insecticides in the cardamom ecosystem and consequent resurgence of the whitefly. The adult is a minute moth like (around 3mm) white insect with a waxy bloom over its body. Eggs are thrust into the leaves and the first instar alone is the mobile nymphal phase. On emergence from the eggs, the first instar crawls on the leaf to fix a suitable feeding site and start feeding on the sap. Further instar nymphs also occur on the same spot except for the movement caused by moulting. The active sucking by the nymphs cause yellowing and discolouration. Nymphs secrete honey dew which fall on the lower leaves and cause growth of sooty mould. Infested plant becomes stunted. For chemical control, spray with 2% neem oil with addition of teepol or triton X-100 @0.05% or acephate 0.07 percent. Never go for insecticide mixtures, or use the same insecticide repeatedly.

Root mealy bug: *Pseudococcus* sp.: Pseudococcidae

Root mealy bugs have assumed pest status in recent years. The eggs are deposited in the root zone, which hatch out and start crawling to locate a suitable feeding site. Later instars may also move, but less actively. There is also a fungal mass sometimes associated with the pest. Desapping by the mealy bugs cause yellowing, wilting and defoliation of the leaves. Plant and plantation health is affected seriously causing reduced yields. Assessment of infestation can also be monitored by the presence and active movement of ants that tend the mealy bugs for their honey dew. The soil may be slightly loosened or excavated around the affected clumps and a solution of Chlorpyriphos at 0.05% strength has to be drenched in the base and the soil put back. This creates a fumigant action in the root zone and also by contact action, causes the death of the mealy bugs. A repeated application may be done after a few weeks if the infestation is not fully contained.

Lacewing bugs: *Stephanitis typicus*: Tingidae

Hemiptera

Adults are small, dull-coloured or whitish bugs with lace-like reticulate wings. The nymphs are black coloured. Both nymphs and adults are seen in colonies on undersurface of leaves feeding on sap and causing discolouration of leaves. These are usually pests in the summer months. Greyish yellow spots develop at the feeding site.Usually no chemicals are recommended. If needed, a spray with less residual chemicals can be given.

Cardamom aphid: *Pentalonia nigronervosa* F. *Caladii* : Aphididae

It causes indirect damage by transmitting the "Katte" or marble mosaic disease. The adults and nymphs are seen in colonies on leaf axils and pseudo stem. Alternate hosts are *Colocasia* sp., *Alocasia* sp. and Banana.

Adult aphids are brownish in colour. They have two forms; the winged and wingless forms. Winged or alate forms have black-veined wings. It lives in colonies within leaf-axils, or tender leaves and at base of the culms at the ground level. They reproduce by giving birth to young ones .This phenomenon in aphids is called vivipary. The females live parthenogenetically. About 35-50 nymphs may be produced by a female during a life span of 27-37 days. A female gives birth to one nymph each within every 3-4 hours. Nymphs undergo four instars of 8-10 days duration. Total life cycle takes 12-15 days for completion. About 30-40 overlapping generations may be completed in a year.

At the time of planting, apply 25 g of phorate or carbofuran to the soil. A second application of 12.5 g in the leaf axil or 25 g in the base at the soil is recommended followed by a third, at 165 days, if attack persists.

Monitor for the incidence of the pest and remove and destroy affected plants immediately.

Scale insects: *Aspidiotus destructor*: Diaspididae

These scales usually infest lower surface of the leaf lamina in colonies. Nymphs and adult females feed by sucking the sap. Infestation leads to yellowing of the leaves in specks on the areas where the scales are present. Scale infestation peaks during summer months. In case of severe infestation spraying with fenthion 0.1% or dimethoate 0.05% or Monocrotophos 0.05% is recommended. In the field, attack by the scale insects are brought under check naturally by the coccinellid predator, *Chilocorus nigritus*.

Shoot fly: *Formosina flavipes*:Chloropidae

This is a pest that attacks the seedlings in the nursery and young tillers. The adults lay cigar shaped eggs between the leaf sheaths. The emerging maggots bore into the central shoots, and cause dead hearts or decay of central spindle.

Avoid excessive nitrogen, provide adequate shade, and spray quinalphos 0.05 percent or apply carbofuran @ one gm a. i /sq.m of nursery or 20 kg/ha.

Spotted grasshopper: ***Aularches miliaris*** **:Acrididae**

This is an occasional pest assuming locust proportions in some years, feeding on shade trees, arecanut, banana and wild plants. Nymphs and adults feed gregariously on the vegetation including cardamom leaves, and produce a frothy secretion between the head and thorax when challenged. Adults are green and beautiful to look at, and the spotted with yellow markings.

In the swarm phase, rake the egg pods and expose the eggs in the soil, collect and destroy the nymphs and adults, and if needed, spray quinalphos 0.05 percent or dichlorvos 0.75 per cent.

Rhizome weevil: ***Prodioctes haematicus*****: Curculionidae**

The adult is a reddish brown weevil, about 12 mm long, with three longitudinal black lines on pronotum and three black marks on each elytron. This is a pest taking about one year for one life cycle, with the adult living for about 7-8 months. They emerge in large numbers with the first monsoon showers and lay single eggs in the cavities in the rhizomes. The grubs come out in around 8-10 days and tunnels into the rhizome. They are fully fed by three months, and then pupate in the soil for about 3 weeks. The feeding by the grubs cause decaying of the rhizomes and they die. Application of chlorpyriphos or other dust formulation in the rhizome region is recommended.

Other Pests

Early capsule borer: ***Lampides elpis*****;** ***Jamides*** **sp. : Lycaenidae**

The larva attacks the pods and flowers.

Galerucid borer :***Thamnuroides cardamom*****: Galerucidae**

The small dark brown grubs with hairs all over the body make tiny circular bore holes on the capsules.

Soft brown scale ***Saissetia coffeae*** **: Diaspididae**

These scales with hard outer cover remain on the plants and cause spot yellowing

Pink leafhopper *Bothrogonia* sp. :Cicadellidae

Nymphs and adults feed on the leaves and cause yellowing and chlorosis. The adults are pinkish wedge shaped adults having black spots on the head and thorax and move sideways usually.

Hairy caterpillars : *Eupterote cardamom*: *E. canaraica, E.fabia and E. testacea* :Eupterotidae (= Bombycidae)

Eupterote cardamomi

Eupterote canaraica

These are large moths pale yellow to brown, with wavy lines and a series of spots on the outer margin, or zig zag lines, or black wavy lines depending on the species.

The larvae are blackish with different designs depending on the species. They are defoliators found attacking shade trees in their initial stages and descending to cardamom through silken threads during the 6th or 7th instars. They are serious only in some years.

Collection and destruction of the caterpillars in their late instars can usually manage the pest.

Root borer: *Hilarographa caminodes*: Yponomeutidae (= plutellidae)

The larvae are root borers, tunneling them and causing yellowing of leaves and weakening of the trees.

Eupterote canaraica

Eupterote cardamomi

Non-insect pests

Root knot nematode: ***Meloidogyne incognita and M.javanica.*** These are sometimes serious pests of cardamom nursery. The attack causes profuse tillering, stunting and weakening of tillers and production of abnormal narrow leaves.

Red spider mite: ***Dolychotetranychus elettariae. D.floridanus*****: Tetranychidae**

These mites attack the leaves and cause yellowing and red lesions.

Chapter - 56

CINNAMON

Cinnamon butterfly: *Chilasa clytia clytia* : Papilionidae

Larvae defoliate plants. The larvae are pests on many crop plants and weeds and are common in the plains. The larvae are jet black at hatching. They develop a series of broad white patches. The adult is a common butterfly rich velvety brown, and spotted with white dots. The hind wings have arrow shaped streaks and crescent shaped spots.

Common blue butterfly : *Graphium sarpedon* and *G. doson* : Papilionidae

Adults are brownish black butterflies. The forewing has transverse band with nine pale emerald green spots for *G.sarpedon* and many marginal spots for *G.doson.* The larvae are initially spiny and smoky, and then become green with longitudinal lateral and lower yellowish lines. Both species are defoliators.

Shoot and leaf webber : *Sorolopha archemedias* ;Tortricidae

The larvae are defoliators .They roll and web the leaves. Continuous infestation reduces the normal growth of plants.

Looper caterpillar : *Thalassodes quadraria* : Geometridae

The adults are small greenish moths and the larvae are brownish loopers that take on the pose of a leaf petiole for escape from predators. The loopers take very odd shapes like dead twigs and consume a lot of foliage, especially of the tender flushing. The new growth is affected.

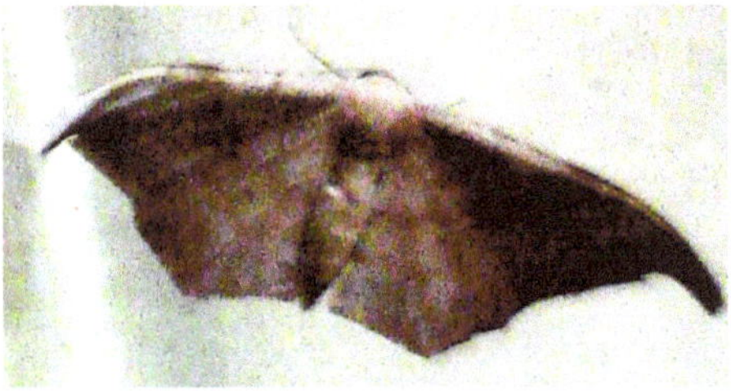

Leaf miner: *Conopomorpha civica, Acrocercops* Sp.near *telestis, Phyllocnistis citrella*: Gracillaridae: Lepidoptera

These are small gracillarid larvae. Due to the feeding, the leaves have blister like patches. Larvae mine tender leaves resulting in drying of infested portions.

Jumping Lice: *Pauropsilla depressa* : Homoptera : Psyllidae

These are gall making insects .The pests not only affect the growth of the plant by making galls on the leaves but also reduce the photosynthetic activities. The nymphs are minute and yellowish .

Pink leafhopper :*Bothrogonia* sp.:Cicadellidae

Nymphs and adults feed on the leaves and cause yellowing and chlorosis. The adults are pinkish wedge shaped adults having black spots on the head and thorax and move sideways usually.

Red borer, *Zeuzera cof-feae*: Zeuzeridae (=Cossidae: Lepidop-tera)

Red borer is a pest which bores into the young stem and also the main shoot It has a wide range of hosts including tea, cinnamon, sandal, cotton, orange, teak and many forest trees.

Larva causes damage by boring the stem or branches to feed on the wood. In early stages of attack, young plants or branches show signs of wilting. Infested part bears one or two holes through which, pellet-like excrement of the larva hangs out and accumulate at the base of the plant. In advanced cases, the branch or the whole plant dries up.

Adult of orange-red borer is a medium sized moth with many black spots and patches on both the wings, but more on the forewings. The male is more whitish while the female is more tanned and also has a long, broad abdomen. Eggs are laid in strings on the barks. Incubation period is 8 to 12 days. Larva is reddish, and hence the name red borer. Usually, the excrement pellets pushed out by the borer is also reddish. After entering through the junction of leaf stalk and twig, larva constructs a tunnel in young plants and starts boring itself down to the root region. Larval and pupal stage together last for about 12 to 24 months. After moth emergence, pupal skin protrudes outside through exit hole.

Red Wax scale: *Ceroplastes rubens*: Coccidae

The peculiar looking scales desap from the leaves and shoots occasionally in both nymphal and adult stage. Usually they are seen on the area near the veins and petioles, but during heavy attack, can spread to the entire lamina.Leaf webber , *Orthaga vitalis*:PyralidaeThe larvae are leaf webbers and attack the shoots especially in the new flush stage.**Lymantrid caterpillars:*Dasychira mendosa* and *Euproctis fraterna*:** These are also defoliators in cinnamon although occasionally.

IPM in cinnamon

1. Cinnamon is almost never chemically protected. Mechanical control of the larvae and other pests are sufficient. If the red borer attack is noticed, Cut and burn red borer infested plant or twigs
2. Systematically monitor the plants for the presence of the borer, especially the pellet like excreta in the soil or on the plant. Spray the bark or trunk with a contact insecticide in and around the infested plants and in endemic attacks. The insecticides will also take care of looper caterpillars and other defoliators.
3. The entomopathogenic fungus *B. bassiana* can be used against the pest on a continued basis.

Chapter - 57

CLOVE

Stem-borer: ***Sahyadrassus malabaricus*****: Hepialidae**

Presence of the pest is understood when chewed tissue is seen at at the col-lar region. The mouth of the tunnel in which the borer feeds show pieces of bark, wood and frass particles held together by silk. The saplings break off at the point of injury. When the main stem is attacked, there is mat-like frass at basal region of stem. Larvae girdle main stem and bores downward killing the tree.

Adult moth is peculiar and hangs vertically down by the support of two pairs of legs. Egg period is for 7-10 days, and larvae live for about 10 months inside the larval tunnels. They pupate for 3-5 weeks.

Management is done by monitoring, and cleaning the base of bush. The hiding larvae inside the tunnels are mechanically killed. Injection of insecticides like dichlorvos @ 2ml/l and plugging the holes can also be practiced.

Scale insects: ***Ceroplastes floridensis , Kilifia accuminata*** **:Coccidae**

The pest is problematic in the seedlings and the young plants where sucking of the sap by the nymphs and adults cause wilting of seedlings and young plants in the nursery. These plants eventually may dry up.

Black scale: *Saissetia nigra;* Green scale: *Lecanium* sp. : Coccidae

Green scale is usually not noticed due to its colour. The nymphs and adults suck the sap from the leaves and causes sooty mould on the honey dew secreted by them.

Hard scale, Masked scale : *Mycetaspis personata* : Diaspididae

These are dome shaped greyish brown scales that get attached to the leaves, petioles and shoots and cause yellowing, discolouration, and dropping of the leaves.

Guava scale : *Chloropulvinaria(pulvinaria) psidii* : Coccidae

The soft scales infest the leaves and tender shoots of guava on the underside. The egg sac can be seen and the mother scale moves off leaving the egg sacs as a trail. (See photo). Desapping in large numbers by the scales cause loss of vigour for the plants.

Jumping Lice :*Pauropsilla depressa*: Psyllidae

These are gall making insects .The pests not only affect the growth of the plant by making galls on the leaves but also reduce the photosynthetic activities. The nymphs are minute and yellowish and cause the galls on the leaves.

Chilli thrips: *Scirtothrips dorsalis*: Thripidae

The adults of this insect are small brownish black with fringed wings. The nymphs are slender, active within the folds and yellowish straw coloured, later acquiring more brownish hue. They lacerate the tissues and feed on the exuding sap. Leaves become crinkled, curled upward and shed. Buds become brittle and drop down. Plants get stunted and bronzed.

IPM in clove

Except the stem borer, affecting the young seedlings and young plants, the clove tree does not get affected by many pests. The attack by the sucking pests like scales, thrips and mealy bugs are not significant.

On few occasions, the psyllid causes galls on the leaf, but this is also not significant. Clove is best grown without insecticides.

Chapter - 58

CORIANDER, CUMIN, FENNEL AND FENUGREEK

Coriander aphid: *Hyadaphis coriandri*: Aphididae

Coriander aphids are yellow-green in color, dusted with grayish wax. They have short, dusky, slightly swollen, siphunculi (or cornicles). The Aphids form dense, damaging colonies on leaves, heads, and stems of their host plants.

In its native land, Florida, USA, the life cycle of the coriander aphid is similar to that of other host-alternating aphids. Overwintering eggs occur on other hosts. A fundatrix (or stem mother) hatches from the egg in the spring. Her offspring are parthenogenetic, and the winged-females are spring migrants. They colonize summer host plants in the family Umbelliferae. During the summer, there are many generations of aphids. All individuals are parthenogenetic females, but they can be either winged or wingless. When the colony becomes crowded, winged individuals (alates) are formed.These emigrate to newer areas and form a new breed on the same type or different hosts. Adults and nymphs feed on plant sap and cause wilting of tender shoots, leaves and flowers and pods. The honey dew secreted by the aphids aid in formation of sooty – mould.

When the pest becomes severe, spray Methyl demeton 25 EC 1ml/l, dimethoate, 30 EC 1 ml/ lit or neem based formulations 2 ml/lit

Green peach aphid: ***Myzus persicae*****: Aphididae**

Leaves get curled and crinkled, and are coated with honeydew and sooty mould. Plants remain stunted. Adult is mostly yellow in colour.

When the pest becomes severe, spray Methyl demeton 25 EC 1ml/l, dimethoate, 30 EC 1 ml/ l or neem based formulations 2 ml/l

Thrips: ***Thrips flavus*****: Thripidae**

Infestation of thrips is generally during hot months with night humidity. The attack by thrips leads to discolouration, bronzing and drying of affected leaves.

Semilooper: ***Thysanoplusia(=plusia) orichalcea*****: Noctuidae**

Bluish green or grey coloured eggs are laid singly on lower surface of leaves. Larvae feed voraciously on leaves and defoliate plants. Add photo. Adults are brownish with golden yellow angulated bands on the forewings.

Seed midge: ***Systole albipennis***

Larvae feed on developing embryo and endosperm of fruits.

Pentatomid bug: ***Dolycoris indicus*****: Pentatomidae**

This is a summer pest. Maximum population of the bug was found during May. Both nymphs and adults suck sap from growing spikes and cause flower drop.

Chapter - 59

GINGER AND TURMERIC

Shoot borer: ***Conogethes punctiferalis*****: Pyraustidae**

The main damage done by the caterpillar of this pest is to bore holes in the pseudostem and to plug it with excreta, causing dead hearts. Larva is pale greenish with pinkish tinge and dark head and has fine hairs on the body with a pro thoracic shield. Adult is a medium sized moth, pale yellowish with small black spots on the fore and hind wings.

Management involves spraying dimethoate or malathion 1.5 ml/lit or phosphamidon 0.75 ml/lit of water. This is mainly intended to manage the adults flying in the crop canopy and the first instar caterpillars that are seen on the outside of the plants and before thsy enter the pseudostem. The spray has to be given at the very first spotting of dead haearts.

Rhizome maggot: *Formosina flavipes, Chalcidomyia atricornis*: Chloropidae

Both the species may be seen as the primary pest causing ginger or turmeric rot or as a secondary attack after the rhizomes have got an infection with the rhizome rot fungus. The maggots tunnel the roots and rhizomes extensively resulting in rotting of rhizomes. Management has to start with drenching of the base of the affected rhizomes with a chemical like chlorpyriphos 5 ml/l or Imidachloprid@5 ml/10 l and slightly raking up the soil and heaping at the base of the plant. A general spray on the crop can be given to kill the roaming adults and to prevent further egg laying.

Rhizome maggots, *Calobata* sp., *Mimegralla coeruleifrons*, Micropezidae;*Eumerus albifrons*:Syrphidae, *Celiphus* sp. : Celyphidae.

The adult flies lay their eggs in the rhizomes. The developing maggots of the dipterans feed inside the developing rhizomes. The rhizomes get rotten and the yield is reduced. Rhizomes get rotten within around two weeks of infestation. The plants become yellow, wither and the pseudo stem is lost. The pupation is also inside the rhizome in the soil. The affected plants show yellowing of the leaves from centre to the outside.

These pests are difficult to control. in the plants that show the yellowing symptoms, the soil above the developing rhizomes has to be scraped to the sides, and insecticides like chlorpyriphos (@5ml/l) solution drenched at the spots of infestation, so as to kill the maggots fully. Application of carbofuran is also practiced, if the yield is for seed purpose or if there is a waiting period of 100-110 days possible before harvest.

Rhizome scale: *Aspidiotus (Aspidiella) hartii*: Diaspididae: Due to the attack, plants become weak, pale and withered in the field and results in shrivelling of rhizomes and buds. Adult is yellow to deep yellow in colour and is seen as sedentary warts on the rhizome. The adults are minute, circular light brownish to grey with a thin pale membrane. The inner side of the scale shows yellowish to deep yellowish, round insects inside.

Management

i. Soil drenching of dimethoate @ 2 ml/l or chlorpyriphos @10ml/l of water

ii. For seed rhizomes, soak in insecticide solution of chlorpyriphos (@5ml/l) or dimethoate 1.75 ml/l or phosalone 1.5 ml/l or dichlorvos 0.5 ml/lit for 15 min. and dry in the shade before storing. Add carbendazim 50 WP 1 g/l to the solution.

Thrips :*Panchaetothrips indicus* :Thripidae

The thrips attack the leaves, which become rolled up, turn pale and gradually dry-up. Adult is with fringed wings. When the symptoms are severe, spray dimethoate @ 1 ml/l.

Lace wing bug: *Stephanitis typicus* :Tingidae

The nymph and adult bugs are gregarious in their habit. They feed by sucking the sap from the undersurface of the leaf causing white specks on the upper surface which later turns yellowish. The population of the lacewing bug is highest during March to May and lowest in December. Adults are brownish black coloured with transparent wings and netted venation. Nymphs are whitish with dark patches. The pest has a wide range of host plants which includes banana, ginger, turmeric, cardamom etc. *Stethoconus praefectus*, a mirid bug is the major predator of the nymph and adult bugs

Leaf roller: *Udaspes folus*: Hesperiidae

The adult is a medium sized brownish black butterfly with a large white stretch on the inside of the hind wing. The forewing has 9 white patches, arranged more towards the periphery of the wing, and with different sizes. Adults lay about 50 spherical pale brown single eggs on the under surface of leaves. The eggs hatch in 4-5 days and complete 5 larval instars in 14-25 days. Caterpillars have smooth green body and a black head with a constriction of the larval body commencing from the thorax. They cause tubular rolls of the leaves inside which they feed. Pupation is inside the roll and adults emerge in about a week. The larvae are important as defoliators of the crop. If the attack is severe, contact insecticides like malathion, quinalphos, carbaryl, dichlorvos, fipronyl or endosulfan can be sprayed

Leaf beetle: *Lema* spp.: Chrysomelidae

This is a small beetle whose larvae have very small legs, a large swollen body and usually carries its excreta over it body. They emerge as adults during the

pre monsoon rains and lay around 100 yellowish, cylindrical eggs on the leaves. Grubs emerge in a week and undergo five moults in a period of 25 days. The grubs have a dull white body with brown streaks and a shining black head. Pupation is in the plant in an odd fluffy cocoon for more than a month. The damage is caused by the adults and grubs which cause defoliation.

Application of insecticides against shoot borer or leaf rollers would suppress the beetle attack also.

Leaf weevil: *Hedychrous rufofaciatus*: Curculionidae

Adults of this weevil are occasionally seen causing shot holes and leaf damage on the margins.

Chapter - 60

LARGE CARDAMOM (AMOMUM)

Grasshoppers: *Mecopoda elongata* :Tettigonidae; *Oxya velox*: Acrididae

These pests occur both in nurseries and plantations. Both adults and nymphs feed on leaves by scrapping and eating outer tissues and margins. The emerging leaves are nibbled and the spot holes appear when the leaf unfurls. In severe conditions the entire leaf is eaten leaving behind only the midrib.

Hairy caterpillars: *Clelea plumbiola*: Zygaenidae

The pests appear during July-October. The caterpillars feed on chlorophyll content in the initial instars as gregarious forms leaving behind only the skeleton of midrib and veins. The infestation sometimes even extends to young flower and sucker buds. The fully -grown larva usually migrates and pupates in soil in membranous cocoon. The adult is small, measuring 1.54-2.0 cm wingspan, light greyish forewings and black hind wing. The adult lays eggs on the under surface of the leaves.

Stem borers: *Glyphipterix* sp.: Glyphipterigidae

The tiny caterpillar bores into the leafy stem at a height of about 5-10 cm above the collar region, and feeds on the central leaf core, causing dead heart. The pest is active during July- September. This is a very common insect causing substantial losses and hence often requires strict control measures. The chemicals recommended can be quinalphos,Triazophos 40 EC 2.5 ml/l WP or fipronyl 5SC 2ml/L.

Capsule borers: *Synoxy* sp.: Scolytidae

This is a serious pest attacking the immature capsules. The grub measures about 1.5-2.0mm in length, bores through the capsule wall and eats away the young soft seed mass, making the capsule hollow. They also pupate inside the capsule.

Leaf hoppers: *Kolla mimica* and *K. Opponens*: Cicadellidae

These leafhoppers sometimes congregate on the under surface of the leaves sucking their sap and cause yellowing.

Thrips: *Rhipiphorothrips cruentatus*:Thripidae

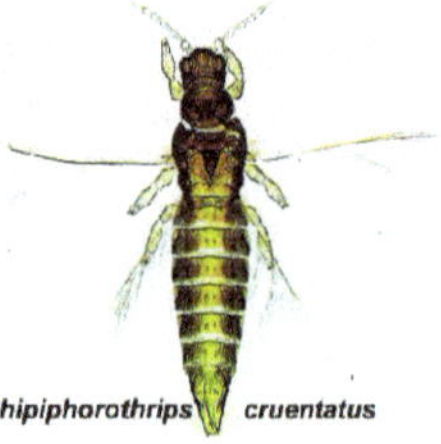

The thrips feed on the leaves and capsules causing corky encrustations and thickenings and cause loss flavour and quality.

Mealy bugs and scale insects: *Paraputo* sp. Pseudococcidae

These are subterranean inhabiting pests, sucking the sap from developing capsules, the actively growing root tips and differentiating buds on mature rhizomes. Their incidence is more severe in drier zones, in plantations with scarce shade and those areas nearby ginger and turmeric fields.

Aphids: ***Pentalonia nigranervosa; Rhopalosiphum maidis*** **: Aphididae**

These pests are known to be the vectors of 'foorkey' virus disease. They suck sap from the tender portions and differentiating buds at collar region. If abundant they can be seen on leafy stems. The affected clumps show the foorkey symptoms. The differentiating spikes and the floral parts are very susceptible to aphid attack. The infestations occur mainly during May-November.

Corn aphids are seen on lower surface of the leaves congregating near the midrib and veins. These are known to be the vectors of mosaic streak or 'chirkey' virus disease. They suck the sap from the leaves and secrete honey dew, which encourages the sooty mould growth on leaves. This is a major pest causing the outbreak of 'chirkey' viral disease.

Chapter - 61

NUTMEG

Scale insects:*Pseudaulacaspis cockerelli, Parasaisettia nigra*:Coccidae

Attack Adults and nymphs feed on plant sap and cause wilting and drying of seedlings and young plants.

Brown soft scale: *Lecanium* sp. *(=Coccus) hesperidum* : Coccidae

Adults are oval, flattened, slightly convex, reddish-brown scale insects. Eggs are produced within the female and hatch into first stage nymphs, or crawlers. The mobile crawler stage has three pairs of legs, is pinkish, oval in shape and has two prominent hairs (setae) arising from the end of the abdomen. Crawlers disperse and find a suitable feeding site.From second instar onwards they resemble adults but are smaller. Winged adult males and females emerge in about a month. Six to seven generations can occur per year .They are dissappers from the shoots of nutmeg, but are minor pests.

Jumping Lice: *Pauropsilla depressa*: Psyllidae

These are gall making insects. The pests not only affect the growth of the plant by making galls on the leaves but also reduce the photosynthetic activities. The nymphs are minute and yellowish and stick to the leaf initially and then get themselves inside the gall.

Chapter - 62

ONION AND GARLIC

Thrips: *Thrips tabaci*: Thripidae

The nymphs and adults of the thrips desap from the narrow spaces between the inner leaf axils. Leaves are faded, with curled tips and shining silver streaks or brown spots. Many minute yellow or white insects are seen moving, generally near the base or middle of leaves and they suck the sap from the plants. Plants are faded and the yield is reduced.

Mite: *Aceria tulipae*: Eriophyidae

These mites desap the leaves and result in malformation of the leaves causing reduced yields. Infested leaves show yellow patches. Plants get stunted.

Onion fly: *Delia antiqua*:Muscidae

This is coming under the family of houseflies. The maggots feed on the developing bulb underground and cause rotting. Plants wither, and attack predisposes to infection by fungus causing soft rot of onion.

Cutworm: *Agrotis ipsilon*: Noctuidae

The Adult has pale brown forewings with a dark purplish brown tinge on the costal margin. The hyaline hind wings have a tinge of brown. The male has a distinct bipectinate antenna.

Caterpillars are nocturnal in habit and hide during the day in cracks and crevices in soil or under debris around plants. At night they come out, cut seedlings near ground level and eat the tender parts. Full-grown caterpillars enter soil and pupate in earthen cocoons. Egg, caterpillar and pupal stages last for 2 to 3, 10 to 30 and 8-10 days, respectively. Total life cycle is 30 to 68 days.

Ground nut earwig: *Anisolabis (Euborellia) annulipes, Anisolabis stalli;* Forficulidae

Due to the attack, the young bulbs show bore holes plugged with excreta, sand particles or discoloured pulps. Bored bulbs are unproductive. The adult is dark brown to black with forceps like caudal cerci and white leg joints. It lays eggs in clusters of 20-100 in soil and sometimes inside damaged pods and hatch in about a week. The five nymphal instars resemble the adults which can live as long as 250 days.

Management

1. Apply malathion 5D or endosulfan 4D or carbaryl 10D at 25 kg/ha prior to sowing in areas where the earwig is endemic.
2. Repeat the soil application of any one of the above dust formulation if the attack persists.

Chapter - 63

PEPPER

Pollu beetle: *Longitarsus nigripennis*:Chrysomelidae

This pest is monophagous to pepper and no alternate host has been reported so far.

Nature of Damage

Adult is a shining blue flea beetle with yellow head and thorax and has thick enlarged femur and tibia. The adult beetle is only the size of a pin head. The infested berries dry up and turn dark in colour. Berries become hollow and crumble when pressed and such hollow berries are called "Pollu" meaning empty. The eggs are laid on the berries @ 1-2 eggs in each hole. Eggs hatch in 5-8 days. The grub period extends for 30-32 days. The emerging grubs feed on the berries of pepper and consume the internal contents. Grub may also eat into spike causing the entire region beyond it to dry up.

Adults damage the leaves by feeding shot holes, on the growing shoot tips and on the surface tissues of tender spikes. Tender leaves of seedlings and young plants suffer severe damage.

When contents of one berry is exhausted the grub moves to the next berry. Thus a grub can damage 3-4 berries, before the grubs fall down and pupate in the soil in a depth of 5.0 - 7.5 cm for 6-7 days.

Incidence of the pest has been recorded at a lower level (5- 10 percent) in the higher altitude of Wynad or above 900 m above MSL and is higher (20-32 percent) in the plains. Estimates of losses due to the pest range from 6-40 per cent.

Management

This is a major pest of black pepper in south India. The attack is more in shaded plantations and hence, shade regulation by pruning the branches of shade trees just before the monsoon is recommended.

In severe incidence, application of insecticides is recommended. Three rounds of insecticides with a rotation of Dimethoate, malathion and quinalphos on a need basis at one month interval may be sprayed. This coincides with spike emergence, berry formation and berry maturity.

Top shoot borer: *Cydia hemidoxa*: Eucosmidae

The top shoot borer is a pest in the nursery and young plantations in the initial years. It is debatable whether this pest is a blessing in disguise, at least in low infestation levels, since attack by the pest initiates development of new shoots from axillary buds.

The adult is a small beautiful moth with a wing span up to 15mm. The basal region of the wing is yellowish and the distal half crimson red with black wavy markings, the forewing being folded roof like in the resting posture. The hind wings are grey. The caterpillar when full fed is 12-14 mm long and is greyish green. The larvae bore into the terminal shoots and eats up the internal contents. The larva is capable of completing its life cycle within a single inter node. There are five larval instars lasting up to 14 days in toto. Pupation is in the larval tunnel extending about 8-10 days. The attack causes the affected terminal tips to dry up. Infestations are heavy during the main flushing season commencing with the onset of monsoon and continues until no more top shoots or terminal shoots are produced. The larvae are parasitized by *Apanletes sp., Euderus* sp., *and Goniozus* sp. *Apanteles cypris* and a mermithid *Hexameris* sp.

The pest is not managed by chemicals in isolation ,since sprays against pollu beetle would manage this pest also. Another reason is its low occurrence and ability to induce axillary shoots in the attacked vines.

Berry gall midge: *Cecidomyia malabarensis*:Cecidomyiidae:

This is an emerging pest both on the leaves and vines. The leaves are damaged in characteristic pattern initially in spots and later yellowing. The vines show galling. The maggots infest the berries at the attachment of the berry to the spike. It causes gall like swelling on the tender stalks and shoots.

The galling of vines can cause reduction in growth of the vines and also reduction in yield. At present levels, chemicals directed against other pests could be sufficient against this gall fly also.

Marginal gall thrips: *Liothrips karnyi*: Thripidae

Thrips attack on the leaf is seen in all pepper tracts, but they can be considered as a minor pest, always available. The attack is initiated on the tender leaves and coincides with the pre-monsoon and monsoon periods, when large number of new leaves are put up. The leaf margins curl downwards and inwards, allowing a concealed feeding site for the thrips. Egg laying of the first brood is on the margins which causes hypertrophy and aided probably by the thrips secretions, folds to the underside, causing the galls. This allows further concealed feeding sites for colony development. Eggs hatch in 6-8 days,

into light brownish thrips that pass through one more nymphal, pre-pupal and pupal instars spanning 13-18 days to become black thrips. The folded margins initially show yellowing, and later blackening and crinkling. The margin becomes malformed, chlorotic and galled. Except in the case of very young vines and even there, only in severe infestations, the pest causes serious concern. As such, even though this is a pest in all pepper tracts, it is not a significant one, since only the leaf is affected, and that too only at the margins.

Management

There are many natural enemies including *Montandoniola moraguesi*, an anthrocorid bug and a predatory thrips *Androthrips flavipes,* a member of Phalaeothripidae that prey upon the juveniles of the thrips. However, insecticidal management specifically for thrips is not necessary. Application of Dimethoate 0.05% on new flushes or Malathion 0.1% or Fenvalerate is recommended, but any of these for pollu beetle and /or top shoot borer as the first spray would manage thrips also.

Scale Insects

Scale: *Lepidosaphes* piperis:Diaspididae;Soft Scale: *Marsipococcus* (=Lecanium) marsupialis: coccidae;

Coconut Scale: *Aspidiotus destructor*: Diaspididae

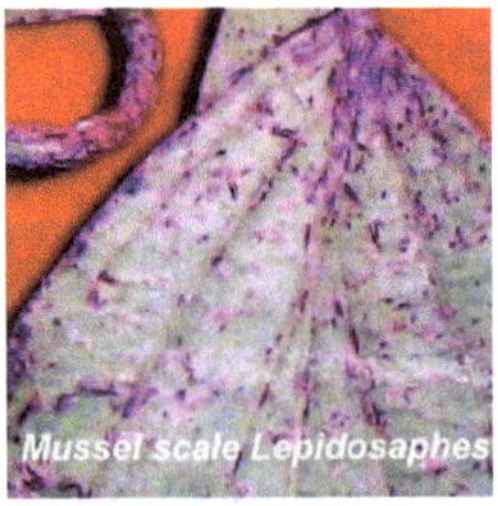
Mussel scale Lepidosaphes

Scale insects, especially the mussel scale, have become serious pests in pepper nurseries and small plants as in bush pepper. The pest is more serious since their infestation is not easily recognized by the farmer due to the sedentary nature and also because of their camouflaging colour.

L. piperis, the mussel scale almost fully covers the main stems of young vines and lateral branches. They are also seen on the undersurface of mature leaves and also cover the berries fully. The main damage to the crops is by the sap sucking by the pests which causes devitalizing of the young vines. As consequence, they dry up. The scales also infest the undersurface of leaves, but are limited to a few leaves only. In high altitudes, they also attack the berries. These berries are ugly to look at close range and of low or little market value. When the berries mature, they can be seen as oval dark grey

scales sticking on the berries in large numbers. Initially they cause yellowing of the berries, but even after drying the berries are crinkled and malformed. This scale can also attack cassava.

Aspidiotus destructor sign the coconut scale, is also seen attacking the leaves in large numbers, causing yellowing and chlorotic patches on the leaves. This is highly Polyphagous and attacks more than 20 economically important crops.

Management of Scale Insects

This assumes importance for mussel scale control especially in nurseries. Here, there is no alternative to chemical control, since drying up of cuttings cannot be allowed. Chlorpyriphos 0.05% or Dimethoate 0.05-0.1% depending on the intensity of attack can be sprayed. It is necessary to apply the insecticide along with teepol or Sandovit @ 5 ml per litre for easy penetration of the waxy layer of the scales, even when systematic insecticides are used, since many of these has an added contact and/ or stomach action also. For managing *A. destructor,* which is mostly on the leaves, the normal sprays against other pests may be sufficient.

Root Mealy bugs/mealy bug: *Planococcus* sp.; *Ferrisia Virgata*: Pseudococcidae

These are pests of pepper that have assumed importance recently. Both the species attack the leaves and species. *Planococcus* is also the root mealy bug, which causes yellowing and drying up of the vines in nursery and in young plantations. The root mealy bug is not easily detected and is hence not easy to manage. Colonies of the root mealy bug infest the basal portion of the stem under the soil and also the roots and extensively desap from the tissues. They do not move actively, even though they are not entirely *Ferrisia Virgata* sedentary. Desapping *Ferrisia Virgata* leads to yellowing of the plants initially and later cause wilting and mortality of vines. Nine species of the mealy bugs have been reported to infest pepper.

During recent years, mealy bug infestation

has become widespread on the berries also causing reduction in quality of the produce. Even during harvest, the body parts remain on the berries, which contaminate the dried produce.

Management

The root mealy bug is a pest that is difficult to manage, since it is below the soil and usually undetected until the vines dry up. When the vines start yellowing, drenching with 0.075% Chlorpyriphos should be done. It is most essential to drench sufficient spray fluid. Addition of surfactants like teepol or Sandovit or soap solution @ 5ml /gm per litre drastically improves efficiency of control.

For berry attack, if soap/surfactant to the third round of spraying is done for the spray against the pollu beetle, mealy bug is also managed.

Vine borers: *Diboma procera;Pterolophia griseovaria; P. annulata*: Cerambycidae

These infest the vines and feed on the central core. The faecal matter and frass is seen as chewed up tissues inside the dried vines. The attack is severe during the summer months, but initiated during the post monsoon itself. The grubs pupate inside the vines and come out, making an exit hole on the vines.

Field sanitation by systematic collection and destruction of the vines is necessary against the pest. Since the grubs are semi saprophytic, leaving the collected vines in the field without burning them does not cause death of the grubs. Trash burning of the dried vines is thus very important to kill them.

Whitefly: *Aleurocanthus piperis; Bemisia tabaci*: Aleyrodidae

These are sometimes seen on the undersurface of leaves in large numbers, but no damage has been observed.

Root Grubs:*Holotrichia fissa*:and *Leucopholis burmeisteri*: Melolonthidae

The 'C' shaped grubs of melolonthids are serious feeders of the root in light soils. They pose a problem in new plantations and nurseries. Drenching the root bases with chlorpyriphos solution effectively manages the grubs

1. Before the start of the rains, collect and remove all dried up plant parts, dead vines and leaves from the

plantation. Heap these on the borders and burn during the evenings to destroy the over wintering life stages of the insects. This will considerably reduce the initial populations during the berry season. Since the build up of heavy population follow a geometric progression, low initial loads cause longer periods to attain pest loads due to higher success of natural biological control, extended life cycle for the initial broods and phenological asynchrony of host tissues and pest biology.

2. Regulation of shade in the plantation by lopping of branches. This increases the temperature in the microclimate, reduces humidity build up and increase sunlight and photosynthesis for the vines. Pest build up for pollu beetle reduces and also reduces disease build up.

3. Supplementary planting of damaged or weak stands with vines of resistant or tolerant sources.

4. Scales and mealy bugs cause heavy devitalisation of the plants, especially in the nursery and during the summer months or non-reproductive phase. Higher temperatures cause loss of vigour for the plants and higher vitality for the pests. Management strategy would involve the last spray against pollu beetle or an additional spray in endemic areas to be beefed up by addition of a surfactant like teepol or soap solution for enhancing the effectiveness of the insecticide spray.

5. It is not necessary to spray different chemicals against pollu beetle and top shoot borer. A single spray during the flushing and spice emergence would normally suppress pollu beetle, top shoot borer and also thrips. Selection of a chemical like Dimethoate with higher residual action and lower LD_{50} for the first spray increases efficiency of management. For more effectiveness, especially due to mealy bugs and scales, addition of teepol/soap solution to the spray fluid could be a standard recommendation.

6. Quinalphos is another good option as a spray in the pollu spray schedule, since this insecticide has been reported to have some fungicidal effect also.

7. The second spray can be with endosulfan or a synthetic pyrethroid like fenvalerate .This would also help in lower damage to parasitoids or other natural enemies.

8. Cassava, Coconut and Banana harbour many scales and mealy bugs, but when these plants are well cared for, the incidence is low. Removal of left over parts of these crops should also be done.

9. The third spray when warranted can be with a neem based product, to reduce the residues in the harvested product.

SECTION – VI: TUBEROSE

Chapter - 64

AROIDS AND YAMS

Aphids: ***Aphis gossypii*** **:Aphididae**

The aphids congregate on growing points and suck the sap. Honey dew secreted by the aphids can cause sooty mould in severe infestation. Attack leads to yellowish speckling and curling of leaves.

Thrips: ***Caliothrips indicus*****: Thripidae**

Characteristic silvery white specks with black dots which are mostly faecal pellets, develop on leaves. Consequently, the leaves dry up.

Leaf-eating caterpillars: ***Spodoptera litura*****: Noctuidae**

The adult is a stout greyish brown moth with wavy white markings and criss-cross lines on the forewings and white hind wings having a brown patch along

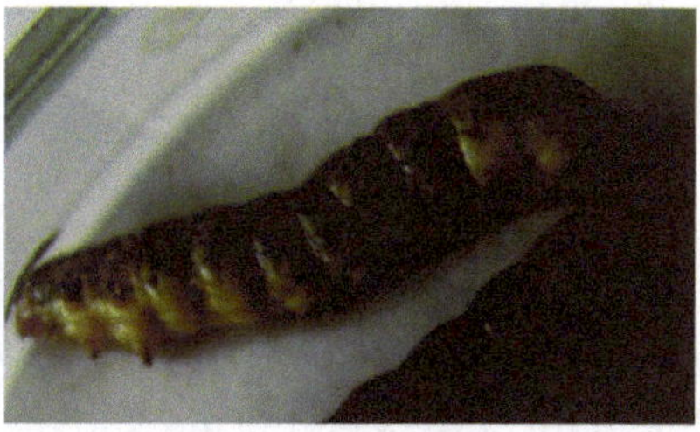

its margin. Young larva scrape leaves on ventral surface, and growing larvae defoliate crops and also feeds on young fruits.

Eggs are laid in masses covered with white hairs on the surface of affected leaves. Eggs hatch in 4-5 days. Larva eats for 14-20 days. They are pale brown with greenish to violet tinge and with stout whitish or creamy bands running the length of the body. Yellow and purplish spots found in sub marginal areas. Faecal pellets are seen on the leaves and on the ground which is the indication of the pest incidence Pupa is seen in the soil for 7-15 days.

Management

- Plough the soil to expose and kill the pupae.
- Grow castor along borders or irrigation channels as indicator or trap crop.
- Flood the field to drive out hibernating larvae.
- Set up light trap at 1/ha or pheromone trap at 15/ha with spheroclin SL lure.
- Destroy egg masses and grown up larvae.
- Apply any one of the following insecticides per ha to control early instar larvae (1st to 3rd instar). Carbaryl 10D 25 kg, fenitrothion 750 ml, carbaryl 50 WP 2 kg, quinalphos 25 EC 2ml/lit 750-1000ml/ha, fenitrothion 625 ml, phenthoate 1250 ml and dichlorvos 750 ml.
- Spray any one of the following per ha to control the 4th to 6thinstar larvae. Chlorpyriphos 2 lit, dichlorvos 1 lit (76 WSC 2 ml/l), quinalphos 25 EC 2ml/lit 1000ml/ha,phenthoate 2 lit.,fenitrothion 625 ml + chlorpyriphos 1.25 lit. in 1000 lit of water. Spray insecticide during evening hrs.

Young caterpillars feed gregariously by scraping on the leaves. Older caterpillars feed on the entire leaf.

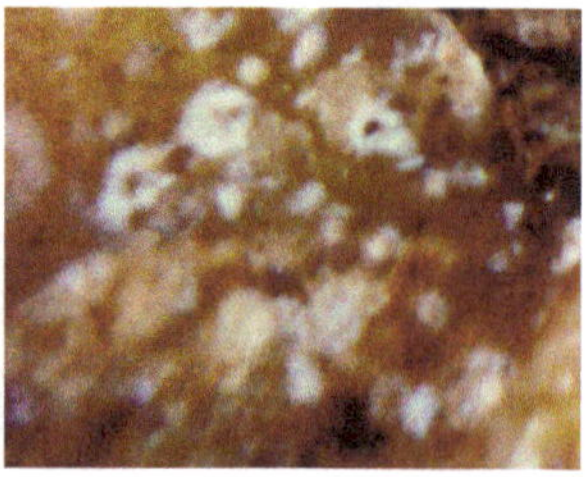

Scale insect: *Aspidiotus (Aspidiella) hartii*: Diaspididae

These are whitish yellow hard scales that desap from aerial parts and tubers. In severe infestation the plant shows withering and drying, reducing vigour and vitality of plants. The infested tubers lose quality and viability.

Arecanut beetle: ***Araecerus fasciculatus*** **: Anthribidae**

This is basically a storage pest. The grubs bore into tubers and convert to powder, causing loss in weight and quality.

Mealybugs: ***Pseudococcus citriculus, Rhizoecus*** **sp.**

Nymphs and adults have a powdery coating and they cover the tuber surface. Sap removal by the mealy bugs lead to shrinkage, poor germination and loss in cooking quality.

Chapter - 65

CASSAVA OR TAPIOCA

Cassava scale : ***Aonidomytilus albus*****: Diaspididae**

White elongated scales appear in colonies on stems. Plants die in the field and under storage. Stem becomes weak and dry, causing side branching to give bushy appearance. Viability of planting material is reduced, resulting in poor establishment.

Termite: ***Odontotermes obesus*** **: Termitidae**

They eat the setts and roots and cause poor establishment and drying of the plants.

White grub: ***Leucopholis coneophora*****: Melolonthidae**

The grubs feed on developing roots and tubers. Attacked plants become pale-yellowish, wither and dry up or have reduced yield. The white grubs or root grubs are more prevalent in sandy or loose soils.

Spider mites: ***Tetranychus cinnabarinus, T. neocaledonicus*****: Tetranychidae**

These are spider mites that colonize on lower surface, causing elongate yellowish streaks, chlorosis, withering and drying of leaves. The photosynthetic efficiency and yield is reduced. Ubiquitous in occurrence.

Eutetranychus orientalis Tetranychidae, *Oligonychus biharensis* Oligonychidae

Both the species infests the upper surface of leaves, causing characteristic bronzing and typical leathery appearance. Such leaves curl from margins and fold into a roll. Heavy incidence causes depletion of chlorophyll content of leaves, photosynthesis and reduces tuber yield.

Thrips: *Retithrips syriacus*: Thripidae

The reddish nymphs and adults lacerate and suck the sap from the leaves. The attacked leaves become pale, and show dirty brown excreta. In severe infestation the leaves roll and dry up. Damage reduces tuber yield.

Spiralling whitefly: *Aleurodicus disperses*: Aleyrodidae

Adults and nymphs are covered with white waxy and powdery scale. They remain in large numbers on the undersurface of the leaves. Damage causes yellowish speckles with crinkling and curling of leaves. Honeydew secretion leads to development of black sooty-mould.

Cotton whitefly: *Bemisia tabaci*: Aleyrodidae

This insect is important as a vector of Indian cassava mosaic disease (ICMD).

Chips borers: *Araecerus fasciculatus,* Anthribidae, *Lasioderma serricorne* Anobidae,*Rhyzopertha dominica* Bostrychidae, *Sitophilus oryzae,* Curculionidae

These are storage pests that bore into dried tapioca and reduce them into a powdery mass in due course. Grubs of these beetles bore the dry bits and consume them. It causes weight and quality loss.

Angumois grain moth: *Sitotroga cerealella*: Gelechidae:

Larvae webbed together the flour, and feed within, causing weight and quality loss.

Red Flour beetle: *Tribolium castaneum* :Tenebrionidae.

The grubs feed on dry cassava, causing weight and quality loss.

IPM in Tapioca

Tapioca tolerates a lot of insect pests. The management practices are mostly for managing the whiteflies, which are mosaic vectors. Light traps, and yellow sticky traps during dusk and dawn are to be used to attract and kill adults. The sticky traps should be cleaned and fresh sticky material like oil or grease should be applied every day morning. For light traps, use CFL tubes which have more rays in the UV range. If pesticide is needed, apply dichlorvos 1ml/ litre along with 5ml teepol. Apply in strips to conserve parasitoids like *Encarsia* spp.

Chapter - 66

CHINESE POTATO OR COLEUS OR KOORKA

Leaf-and shoot folders: ***Hymenia recurvalis*** **: Pyraustidae**

Pycnarmon cribrata;Phostria piasusalis

Hymenia

Pycnarmon

The larvae web together the leaves and feed within. Folding and webbing of leaves are noticed. In heavy attack, the leaf area is reduced considerably

Stem borer: *Nupersha vexator* :Cerambycidae

The adult is a longicorn beetle with a black patch at the posterior end of the elytra. They lay the eggs on tender stems, and the grubs bore into the shoots.

They pupate inside the tunnels and the infestation causes yellowing of the leaves, drying of terminal shoots and frass emerging out of the broken stems.

Root-knot nematode: *Meloidogyne incognita* : Heteroderidae

This is an important pest of Coleus (Chinese potato). Heavy incidence by the nematodes suppresses root growth and cause gall formation and poor growth. Tubers are malformed, hypertrophied and undersized. Both yield and quality of tubers is reduced.

Pest management

Normally, the most important management is to use nematode free cuttings for planting. The cuttings should also be dipped in a solution of an insecticide like chlorpyriphos 0.1 % or dipped in a mud slurry containing carbofuran or phorate overnight and then used for planting. This would also reduce the leaf roller attack for about 45 days. The leaf rollers when severe, can be managed by spraying any contact insecticide like quinalphos or malathion.

Chapter - 67

POTATO

Aphids: ***Myzus persicae, Lipaphis erysimi*** **and** ***Brevicoryne brassicae*** **Aphididae**

Aphids suck the sap from potato foliage, causing wrinkling, cupping and downward curling of young leaves. The older leaves will become yellow, wilt and fall off. They also excrete honeydew on which sooty mould develops covering affected parts with a thin superficial black coating that hinders photosynthetic activity of leaves resulting in stunted growth of plants. In addition, they also act as vectors, for transmitting potato leaf roll virus. Major damage by *M.persicae* and *A. gossypii* is through transmission of potato leaf roll virus.

To control aphids, spray 0.05% malathion, quinalphos, dimethoate or oxy-demeton methyl. Repeat the spraying, if necessary, after 10 to 12 days.

Leafhoppers: ***Empoasca kerri, Amrasca devastans (biguttula biguttula)*****: Cicadellidae**

Nymphs and adults suck sap from veins on ventral surface of leaves. As a result tips of affected leaves become brown, turn upwards and get dried.

During severe infestation brown spots appear on leaf margins, which also get rolled up and subsequently die away, which produces hopper-burn, stunting and crinkling of leaves. The leaves show a burnt appearance.

Adults are wedge-shaped; yellowish-green in colour. A female lays 25 to 60 eggs in 25 to 30 days. Incubation period is 6 to 11 days; nymphal period is for about 25 days and adult longevity is about 15 days for males and up to 90 days for females. Pre-oviposition period varies between 6 and 16 days. If the attack is severe, spray Acephate 75WP one g/l, Imidachloprid20 EC 0.5-one ml/L, Thiacloprid 0.5-one ml/L, Quinalphos 25 EC 2ml/lit or Hostathion 40 EC 2.5 ml/l

Potato tuber moth: *Phthorimaea operculella*:Gelechiidae

This is a major pest of potato in the field and in storage. It is a cosmopolitan pest, found in warmer countries. Larvae of the moth make mines in tender leaves, petiole and stem. They also bore into tubers and form irregular silk lined galleries filled with excreta. This leads to loss of leaf tissue, death of growing points and weakening or breaking stems. Saprophytes grow in such damaged tubers and causes rotting of potato tubers. The alternate hosts include Tomato, Tobacco, Brinjal, Potato, Sugar beet and Solanaceous weeds.

The adults are nocturnal in habit. With the onset of winter, moths fly from godowns to fields and lay eggs singly, near eyes of exposed tubers and on ventral surface of leaves. A single female lays 150 to 250 eggs. Adult is a small moth with silvery body. Fore wing is greyish-brown with minute dark spots and fringes of hairs. Hind wings are dirty white. Eggs are minute, oval and yellowish in colour. Full-grown caterpillars are white, yellow to pale greenish in colour. Pupation takes place in rough silken cocoons. Egg, larval and pupal periods last for 4, 14 and 7 days respectively with about 6 generations in a year.

IPM

1. Select healthy tubers, avoid shallow planting of tubers and plant them to a depth of 10-15 cm.
2. Do earthing-up at 60 days after planting to avoid female moths egg

laying on the exposed tubers

3. Install pheromone traps in the field at 20/ha. Keep pheromone traps in godowns also.
4. Release egg-larval parasitoid, *Chelonus blackburni* @ 30000/ha twice, 40 and 70 days after planting
5. Treat seed tubers with carbaryl 10 % or fenvalerate dust @ 1 kg/100 kg of tubers

Shoot borer: *Leucinodes orbonalis* :Pyraustidae

It is a major pest of brinjal, but also bore into shoots of potato plants. Affected shoots wilt and drop down.

Adults are medium sized moths with whitish forewing having brown patches and dots and an apical blackish patch. The hind wings are shining silvery white with black dots. Larva is pink in color, 15-18 mm long when fully grown. Pupation is in the leaf folds. Mechanical removal and destruction of all infested shoots by burning or deep burying is recommended.

Bihar hairy caterpillar: *Spilosoma obliqua* :Arctiidae

The young larvae scrape the leaves in a group and skeletonise them. Later instars feed fully on the leaves. Eggs are laid in clusters on leaves. On hatching, caterpillars feed gregariously for some time,and later as they grow, disperse and feed voraciously on older leaves.

Larva is stout 25-40 mm long, with seven orange coloured broad transverse bands having tufts of yellow hairs that are dark at both ends. Adult is crimson

coloured with black dots and black antennae, pectinate in male and filiform in female. Wings have black spots.

Cutworms: ***Agrotis ipsilon, A. segetum, Xestia C. nigrum and Peridroma saucia*** **:Noctuidae**

Young larva feeds on tender foliage and grown up larva cuts the stem at collar region. The seedlings and green matter are then eaten during night. They remain at the base of the plant in crevices during the day.

Moths appear after dusk, mate and lay eggs on ventral surface of leaves or moist soil. Freshly ploughed fields are preferred for oviposition. A female lays 300 to 450 eggs in 10 to 15 clusters. Eggs are globular in shape, ribbed and whitish in colour. Tiny caterpillars feed gregariously on foliage for a few days and then enter into soil. Caterpillars are nocturnal in habit and hide during day in cracks and crevices in soil or under debris around plants. At night they come out, cut seedlings near ground level and eat tender parts. Full-grown caterpillars enter soil and pupate in earthen cocoons. Egg, caterpillar and pupal stages last for 5 to 13, 10 to 25 and up to 30 days, respectively. Total life cycle can be from 30 to 68 days.

The description of the common species are as follows

Agrotis segetum**: Common cutworm:** The fore wing is grey with peg and spot like marking and hind wing is white with brown tinge. The Male has bipectinate antenna and female has filiform antenna.The larva is black coloured with brown head. Triangular spots are seen at the lateral side.

Agrotis ipsilon**: black cutworm:** Fore wing is pale brown with dark purplish brown along costal end. Hind wing is dull white. Male has bipectinate antenna. The larvae are black with pale brown head and mid-dorsal stripes. Triangular spots at the lateral region.

Xestia C. nigrum **: spotted cutworm:** Reddish brown fore wing with concave sunken pale area. Hind wing is dull brown. Brownish larva with series of black markings on lateral area.

Peridroma saucia**:Variegated cutworm.** Reddish brown fore wing with dark brown margin. The larva is light brown with 4-7 yellowish markings on mid-dorsal line.

Managment of cutworms.

Cutworms are an occasional malady.

1. Plough the soil during summer months and expose larvae and pupae for predators to feed on them.
2. Install light traps to attract adult moths. Use pheromone traps at 20/ha to monitor and attract male moths.
3. If the plants are attacked, drench collar region of plants in evening hours with chlorpyriphos or endosulfan 2 ml/l.
4. In endemic areas, apply NSKE 5%, endosulfan 1 lit/ha or chlorpyriphos 1 l/ha or neem oil 5 l/ha. Use high volume sprayer and focus nozzle at the collar region.

Ash weevils: *Myllocerus subfasciatus*: F. Curculionidae

M. subfasciatus are brown weevils. The adults and grubs cause damage. Young larvae (grubs) feed on the roots of plants. The adult weevils cut holes on the leaf or eat the margins of leaves in a characteristic way. The grubs feed the roots of plants and cause wilting and drying of plants. But these weevils are generally of minor importance, due to low populations of both the grubs and adults, and do not require specific chemical management.

The adult weevils lay about 300-500 eggs in the soil over a period of three months. The grubs hatch out in a week and start feeding on the roots of the hosts. The grub period is around 60-75 days. The full grown grubs pupate in earthen cells for 10-12 days. The emerging adults fly out and start feeding on the leaves of the hosts. They are polyphagous and can feed on a variety of crop plants and other vegetation.

Management

In cases of considerable damage to the roots in a plantation, the root zone can be drenched with 0.01 % Chlorpyriphos or rake up a dust formulation of Chlorpyriphos or any other dust formulation in the soil. The adults can be warded off or killed by application of contact insecticides like chlorpyriphos 0.05%, Malathion 0.05 % or Quinalphos 0.05% or synthetic pyrethroids at 0.01-0.05 % depending on the chemical nature of the pyrethroid.

White grubs: *Holotrichia excisa, H. repetita, H. notaticollis* (Melolonthinae); *Anomala communis, A. nathani* : Rutelinae

They can also attack groundnut and sugarcane. Grubs feed on roots and tubers; Adults feed on foliage during night.

These are all polyphagous pests. Eggs are laid in the soil near host plants. On hatching, grubs feed on developing roots and tubers of potato as well as other grasses growing around. When full-fed, grubs over winter deep down in the soil. Grubs are C shaped with orange head.

Adults emerge as soon as the soil gets wetted by the onset of monsoon. Adults feed on foliage during night and damage is more during autumn. *Holotrichia* adults are brown beetles with pale thorax. *Anomala* adults are smaller than *Holotrichia* and pale-yellowish beetles.

IPM

1. Do summer ploughing to expose pupae and adults
2. Dust carbaryl 10% or quinalphos 5% D at 25 kg/ha 10 days after first summer rains
3. Install light traps at dusk in the first showers. Set up bonfires by burning trash and dried leaves during dusk to attract and kill adults.
4. Do pre sowing soil application of entomogenous fungus *Metarhizium anisopliae* 1 × 10 FIB/g during May @ 20 kg/ha and incorporate into the soil
5. In endemic areas, apply phorate 10G at 25 Kg/ha during autumn (August-October).

Other Pests

1. Hadda Beetles: *Epilachna dodecastigma*, *Henosepilachna vigintioctopunctata*: Coccinellidae

Both adult and grubs scrape the lower epidermis of leaves in characteristic manner leaving behind riddled areas. The leaves give a grazed appearance. In severe infestation all leaves may be eaten off leaving only the veins intact (Skeletonization), with the leaves or whole plants withering.

Grub

- Grub: Yellowish in color, stout with spines all over the body. Adult: Spherical pale brown and mottled with black spots. *E. vigintioctopunctata* has 14 spots on each elytron. *E. dodecastigma* has six spots on each elytron.
- *Henosepilachna vigintioctopunctata*

Adult

Adults lay 120-460 eggs/female. They are cigar shaped, yellow and are laid in clusters on lower leaf surface. The eggs hatch in 2-4 days .The grub period is 10-35 days. They are yellowish bearing six rows of longitudinal spines with terminal branching. Pupa occupies 5-6 days and is seen on the stem or leaves. Total life period occupies 20-35 days with 7-10 generations/year.

For managing the pest, systematically collect and destroy leaves with egg masses, grubs and pupae and also adult beetles. Spray carbaryl 50% WP 4 g/ lit or endosulfan 35 EC 2 ml / lit or malathion 50 EC 2 ml /l when populations explode. Diflubenzuron or Buprofezin also works well on the beetles

Green leaf Beetle: *Chalaenosoma metallicum*: (Chrysomelidae: Coleoptera)

These flea beetles feed on potato leaves and cause defoliation in South India. These are green coloured beetles and adults are the damaging stage.

Tussock moth: *Dasychira mendosa* :Lymantriidae

This is a polyphagous pest, attacking a number of fruit trees, potato, cauliflower etc. It is a gregarious feeder and causes defoliation of leaves. Full-grown caterpillar is bluish-brown to blackish in colour having red stripes on its head. Body is densely hairy and has compact tufts of long hair on certain abdominal segments with hair pencils on the front and tufts of hairs on thorax and abdomen.

Cotton Whitefly *Bemisia tabaci* and glasshouse whitefly *Trialeurodes vaporariorum*: Aleyrodidae

Both are polyphagous pests. These tiny insects suck sap from ventral surface of leaves and devitalize plants. In addition, these insects also act as vectors for transmitting virus diseases and as such are of great potential importance. Attack causes yellowing and wilting of leaves.

Thrips: *Selenothrips indicus*:Thripidae

These are tiny, slender, fragile insects. Adults have fringed wings. Both nymphs and adults scrape epidermal tissues of leaves usually near tips and rasp oozing sap. Affected leaf tips get curled and dry up. Spray 0.2% carbaryl or Imidachloprid20 EC2.5-3ml/10 L, Thiacloprid 0.5-one ml/L, to check pest population.

Mites: *Hemitarsonemus latus*: Tarsonemidae

The damage starts from the apex leaf and goes downward. The leaves roll inward, showing oily spots on upper surfaces. The lower surface gets sooty appearance. The leaves ultimately become bronze in colour and wither. The plants become stunted and finally die. Heavy mite infestation might cause 50% loss in yield. It is a major pest in Deccan plateau.

Golden cyst nematode: *Globodera pallida, G.rostochiensis* Heteroderidae

Nematodes are silent killers. Their population has to be in millions to cause appreciable damage to any crop. But these microscopic creatures do reproduce in millions and cause substantial damage in many crops, potato being one. They are also difficult to control since most of these are in the soil, and are not amenable to spraying. Initially, the growth of the plants is retarded and leaves show wilting during hot day and recover in the evenings. Later, the plants become stunted, leaves become yellow and start wilting in patches. The underground symptoms include poor development of roots, and reduction in the number and size of tubers. This is one of the most important pests of potato. Insecticidal treatment of seed tubers, soil solarisation and application of systemic insecticides like Phorate and Aldicarb @20kg/ha, are necessitated in severe attack.

IPM in Potato

Potato IPM starts from planting itself. Avoid growing potato and brinjal in nematode infested field and follow a rotation of once in three years. Seeds should be from nematode free area and avoid irrigation water from infested field to new fields. Sprinkler system during day time would cause cut worms to come out and to be eaten by the birds during day time. Aldicarb 10 G @ 20kg/hectare can be applied in heavily infested fields.

Summer ploughing of the soil exposes larvae and pupae of cutworms to be eaten by birds. This also exposes the grubs of cockchafer beetles for predation. Seed tubers should be free from life stages of potato tuber moth and should be planted at 15cm depth to avoid fresh oviposition. Cover the top layers of the seed tubers with lantana or eucalyptus branches to repel the ovipositing adults.

From the nursery stage itself, set up monitoring traps for the different pests. Pheromone traps against potato tuber moth and cutworms should be installed at the rate of 5/8 per acre. Against white flies, aphids and leaf hoppers set up yellow sticky traps for monitoring. Mechanically remove the nucleus infestations of whitefly, aphid, leafhoppers, epilachna beetles and leaf eating caterpillars. Apply insecticides as spot application at the foci of the infestation to prevent spread to other areas. Pest monitoring should be continuous until the harvest of the crops. Set up light traps or bonfires during dusk to attract and kill adults of cut worms and moths.

Release egg larval parasitoid *Chelonus blackburnii*@12000/acre twice, on 40 and 70 days after planting.

Pest Control using Insecticide

Against cut worms

Apply NSKE 5%, endosulfan 35 EC 400 ml/acre or chlorpyriphos 20EC 1 lit/acre or neem oil 2 lit/ac @500 litre of spray fluid/ha.; use high volume sprayer and focus the nozzle at the collar region. Apply the insecticides in the evening hours.

Whitefly

Spray Quinalphos 400ml or monocrotophos 350 ml or chlorpyriphos 500 ml/ acre.

Epilachna beetles

Carbaryl 50WP 800 g, endosulfan 35 EC 600 ml/acre (450 l of Water.)

Leaf eating caterpillars

Endosulfan 35EC 600 ml/acre (450 l water). For tuber moth, a dose of 1000ml quinalphos or endosulfan may be needed.

Chapter - 68

SWEET POTATO

Sweet potato weevil: *Cylas formicarius*: Apionidae

sweet potato weevil grubs

It is a specific pest of sweet potato and the most important one. Adult weevils are ant-like, slender bodied having elongated snout-like bluish-brown head with geniculate antenna, bright red thorax and leg tips and bluish-brown elytra. Females make small cavities on the tubers or stems and lay eggs singly. Each female lays 180-200 eggs. Grubs are elongate, legless, pale-yellowish white in colour with smaller brown head. Pupation takes place in larval tunnels in the tuber. Incubation, grub and pupal stages last for 5 to 8, 14 to 17 and 4 to 5 days respectively. Life cycle is completed in 4-5 weeks and adults live for 95-110 days. The grubs bore into stems, tunnel inside and feed on soft tissues. Grubs and adults bore into tubers both in field and storage godowns. Affected tubers develop dark patches, which later start rotting. Feeding holes and adult emergence holes can be seen on vines and tubers. In severe infestation the collar region of the plants shows unusual thickening due to proliferation and hypertrophy of tissues causing a characteristic terpenoid odour. Even the slightly damaged tubers are unsuitable for cooking and consumption. Hence yield loss in practice may come to 100% in many cases.

This pest is disseminated from field to field through infested vines and is carried over from season to season by breeding in damaged tubers left in the fields and through infested seed tubers.

IPM

1. Use pest free planting materials. Dip planting materials in chlorpyriphos at 2.5 ml/litre and keep overnight before planting.
2. Remove previous crop residues and alternate host, *Ipomoea* sp. and destroy them
3. Use resistant varieties for planting.
4. Drench soil with chlorpyriphos 0.05% or fenthion 0.05%. Then rake up soil and earth up at 30 and if necessary, also at 60 days after planting in endemic areas.
5. Harvest immediately after maturity and destroy unwanted crop residues to prevent the buildup of the pest.
6. Install pheromone traps to attract and kill the adults.
7. In godowns, treat outer side of bags containing tubers with malathion 5% or carbaryl 5 % D.

Sphingid caterpillar: *Agrius (Herse) convolvuli* :Sphingidae

It is commonly called as hornworm due to the horn like process at the back of the abdomen or as giant hawk moth due to the appearance of the adult moth. There are many alternate hosts for this pest including fruit trees, legumes and vegetables.

They are active during monsoon season. Moths are giant, stout hawk moths, pale grey in colour having pale grey wings with transverse violet bands on abdomen. Females lay conspicuous seed-like shiny eggs singly on the tender parts of plant.

Eggs are sub-spherical in shape. Full-grown caterpillars are robust, green to dark brown in colour with reddish or yellow stripes on sides and a curved horn like process at the anal end. Caterpillars can grow upto 10 cm in length and feed voraciously on leaves. Pupae are reddish-brown in colour and pupation takes place in soil. Incubation, caterpillar and pupal stages last for 5-10, 14-21, 7-11 days respectively. A complete life cycle occupies 4 to 5 weeks. Monitoring and mechanical destruction of the caterpillars alone is required against the pest.

Vine-borer: *Omphisa anastomosalis*: Pyraustidae

Whitish stout caterpillar of this pest bores into vines in South India. The caterpillars bore into the vines, and may kill the vines. Wilting of vines and terminal drying are noticed.Pupation takes place within larval tunnels. Moth has straw-coloured wings with wavy markings on them

Tortoise beetles: *Aspidomorpha miliaris, Cassida (Metriona) circumdata, Chirida bipunctata (c.sexnotata)*: Cassididae

These metallic coloured beetles are active during monsoon. Eggs are laid on ventral leaf surface. Grubs are nocturnal in habit and feed on epidermal tissues of leaves and skeletonize them. Late instar grubs and adults gnaw holes in leaf lamina. Grubs are green, flat with anal projection. Pupation takes place in ventral surface of leaves. Even though there are many species, they are only minor pests. Eggs are laid on the underside of leaves and get hatched in a week's time. The grubs are flat with spiny processes covering their body. Dried excreta are seen on the back attached in masses to the curved tip of the tail. Life cycle is completed in 28 to 36 days.

***Aspidomorpha miliaris*:** These are the largest among the beetles. Broad oval shaped, brownish-red in colour with black dots on the elytra. There are four large dots on all the four sides .

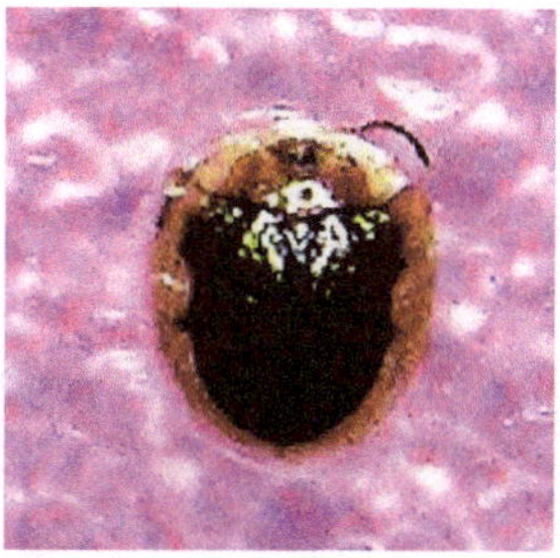

***Cassida (Metriona)circumdata* :** Broad oval shaped, greenish-yellow in colour with green crescent markings on elytra

***Chirida bipunctata (=C.sexnotata?)*:** Small metallic green in colour with six black spots on elytra

Hairy caterpillar: *Creatonotus gangis*: Arctiidae

It is a polyphagous pest, appears in July and continues till November. Adults have shiny black head. Fore wings are straw coloured with pinkish tinge. Hind wings are whitish with few black dots at the margin. A female lays 285 to 695 eggs in clusters arranged in rows closely set together. Eggs are round in shape and shiny creamy-yellow in colour. Incubation period is 4-5 days. Caterpillars are cylindrical, slightly tapering posteriorly and dark violet to black in colour. Meso and Meta thorax are light golden-yellow and head black, hairy with characteristic yellow stripe dorsally. Larval period is about 20-30 days and pupal period 5-7 days. This is also a minor pest.

Blue pansy: *Precis orithya*: Nymphalidae

The adult is a medium-sized bluish butterfly with only two pairs of functional legs. More than half of fore wings are velvety black. Hind wings are blue shaped with velvety black towards the base. The wing pattern thus resembles the pansy flower. It is beneficial also, as it feeds on *Striga*, a weed parasite on sugarcane roots. Incubation, larval, pupal periods, adult longevity and life cycle last for 3, 14 to 16, 4 to 8, 3 to 7 and 27 to 29 days respectively.

For managing the pest, after harvesting, give deep ploughing and flood infested fields to kill pupae and to prevent carry-over of pests. Also, Collect and destroy egg clusters and leaves bearing caterpillars to prevent population build-up. Spray dichlorvos 0.05% or endosulfan 0.05% if the attack is severe.

Leaf folder: *Brachmia convolvuli*: Gelechiidae

It is regarded as a pest of several *Ipomoea* species. Adult is a small slender moth (Microlepidopteran) and greyish-brown in colour. It lays eggs in small groups at the base of radiating veins on the underside of leaves. They hatch in three days. Larva feeds on leaves and becomes full grown in 14 days undergoing five instars.

First and second instars feed on surface tissues of leaves sheltered under thin webbings. Later on they fold single leaves longitudinally and feed on green tissues from within. Attacked leaves dry up. A single larva destroys several leaves. Full-grown larva is slightly flattened and tapering towards both ends. Head is reddish-brown, glossy and flattened; thorax and two abdominal segments are velvety black and other segments are yellowish-white with a velvety black band. It pupates within leaf fold for 7 days.

Collect and destroy egg clusters, leaf folds along with larvae and pupae. Spray dichlorvos 0.5% or endosulfan 0.5% or carbaryl 0.1% or phosalone 0.5% to control widespread infestation

Spiny beetle: *Oncocephala tuberculata*: Hispidae

It occurs commonly in South India. Adult is a small brownish, spiny beetle with blunt projections all over body. It has a life up to 50 days. Eggs are thrust inside leaf tissues. Larva mines the leaf and causes it to wither. When full grown, it constructs a short tunnel in a healthy leaf in which it pupates. Eggs, larval and pupal stages last for 7-10, 17-23,and 8-15 days respectively. Adult also injures the leaves by feeding on them.

Brown looper: *Hyposidra successaria*: Geometridae:

Larvae feed on leaves. Its egg, larval and pupal periods is 4-5, 18-29 and 8-9 days respectively.

Sweet potato hopper: *Exitianus indicus* Cicadellidae

Both nymphs and adults suck sap from leaves and tender shoots, but damage caused is negligible. Adults are active, slender, white leaf hoppers with head, thorax and scutellum greenish in colour.

Fig bug: *Riptortus linearis*:Coreidae

Nymphs and adults infest and damage tender shoots. Adults are elongated and dark brown bugs.They are sap feeders.

Lygaeid bug: *Graptosethus servus*:Lygaeidae

These are greyish-black bugs. Adults and nymphs suck cell sap from tender leaves and devitalize them.

Mealy bugs: *Geococcus coffeae*:Pseudococcidae

Crawlers and adults infest tender roots and tubers. Damage is more severe when slightly infested tubers are stored. Under normal storage conditions, mealy bug multiplies rapidly and stored tubers get thickly covered with mealy growth and become shriveled due to loss of sap. The damage is perpetuated by use of infested tubers as seed. Tubers from infested fields should not be used. Dip the tubers overnight in 0.05% Chlorpyripos or phenthoate solution just before planting.

IPM for Sweet potato

1. After every cropping season remove all crop residues and also destroy the left over nursery.
2. Remove and destroy alternate hosts (*Ipomoea* spp.) of sweet potato weevil.
3. Use planting materials without any life stages of the pests.
4. Seed tubers or vines should be dipped in fenthion 0.05%, fenitrothion 0.05% or chlorpyriphos 0.05%.
5. At 50 days after planting, earth up the ridges to prevent oviposition by adults of sweet potato weevil and for removal of weeds.
6. Tuber traps using 100G bits of sweet potato can be kept at 5m intervals in the evening at 50 to 80 DAP. Collect and destroy the adult weevils every day morning.
7. Do not delay harvesting.
8. In heavy infestation drench the soil with Chlorpyriphos 0.05% at 65DAP.
9. Collect and destroy adults of tortoise beetles with sweep nets.
10. Sphingid caterpillars can be handpicked and destroyed.

SECTION VII - VEGETABLES

Chapter - 69

AMARANTHUS AND SPINACH

Amaranthus stem weevil: *Hypolixus truncatulus* :Curculionidae

It is an important pest of amaranthus and is widely distributed in India. It attacks both wild and cultivated amaranthus, especially the leafy varieties with large leaves. Hence, management involves destruction of unwanted, weedy wild amaranthus plants, to reduce the breeding place for the weevil. Adults are ash-grey in colour, with elbowed antennae and brown elytra. Females lay eggs singly in holes made on the stem, branches, petioles or midribs. Eggs are smooth, oval and pale yellow in colour. Incubation period is 3 to 8 days. A female lays 30-40 eggs. Grubs bore into stems and feed on the internal region, making irregular zigzag tunnels, which are filled with excreta. A single stem can have as much as 17-20 grubs in it. Grubs are stout, curved, legless and white in colour. Grub stage lasts for 15 - 25 days. Pupation is on the stem, usually at ground level or at the base of leaf axils . On emergence, they remain inside the stems for 5 to 6 days, then cut the epidermal membrane and emerge out. Attack causes stunting of plants, twisting and swelling of branches and stems and suppression of shoot and leaf production. The affected stems become weak and split longitudinally. The plants get desiccated and dry up completely. Full-fed grubs form a greyish-brown hard compact gall like chamber. Adult feeds on tender leaves of stem and makes

circular holes in stems, branches and mid-ribs.

IPM

1. Collect and destroy wild amaranthus hosts in the vicinity of cultivated crop.
2. Collect and destroy affected plant parts along with grubs and adults
3. If the attack is severe, a spray of dichlorvos 0.75 ml/l can be given. This has to be after the harvest of the produce. Prolong the next harvest as much as possible.

Amaranthus caterpillar or webber: *Spoladea (Hymenia) recurvalis*: Pyralidae

It is a destructive pest often observed damaging amaranthus in tropical and subtropical regions including Africa, Asia and Australia. In the Indian sub-continent it is found all the year round. Besides amaranthus, it is also a pest of grasslands, pastures and also beans, coleus, melons and spinach.

In the initial life of the larvae, they scrape the epidermal and palisade tissues of leaves. Later on, they web the leaves with silken threads and cause drying of webbed leaves.

Adult is a brownish black moth about 2.3 cm wing span, with white wavy white markings on wings. Eggs are spherical in shape and snow-white in colour and are laid singly or in batches of 2 to 5, in grooves of leaf veins. A female lays 50 to 80 eggs during its lifetime. Caterpillars are greenish in colour with white lines and black crescents on thorax below lateral line and web leaves with silvery silken threads secreted by them and feed within. Gradually webbed leaves become skeletonised and dry up. Many larvae can be seen on the leaves. When full-fed, caterpillars pupate among the leaves. Incubation, caterpillar and pupal periods last for 3 to 4, 12 to 16 and 8 to 10 days respectively. Life cycle is completed in 3 to 4 weeks. If the attack is severe, a spray of dichlorvos 0.75 ml/l or malathion 1ml/l can be given. This has to be after the harvest of the produce. Prolong the next harvest as much as possible.

Bean leaf roller : *Lamprosema indicata*: Pyralidae

The young larvae are gregarious green caterpillars that fold leaves, feed inside and skeletonise them.

Leaf webber: *Psara basalis*:Pyraustidae

The moth is small and delicate about 2 cm in wing span and has yellow wings with faint black dots on them. Their habits, symptoms of damage, life history including management are similar to that of *Hymenia recurvalis*. Damage is also similar, the larvae being leaf webbers. Full fed caterpillars are greenish in colour. A spray as in the case of hymenia may be made if the pest becomes serious

Leaf webber: *Eretmocera impactella* : Heliodinidae

It is a sporadic pest and is widely distributed in the Indian sub-continent. The larvae web the leaves with silken threads and feed from within. Adult moths are small, blackish with prominent yellow spots on fore wings. Eggs are laid on leaves or on top shoots. Caterpillars web leaves with white silken threads and remain hidden in folds feeding from inside. Full-grown caterpillars are cylindrical, brownish-yellow to brownish-grey in colour with a broad sub median dark stripe and black tubercles bearing several divergent longitudinal hairs.

Pupation takes place in white silken cocoons, which remain attached to the leaves. Pupae are long and brown in colour. Life cycle is completed in 3 to 4 weeks. Manage as for other leaf webbers.

Scale: ***Coccus hesperidum*** **:Coccidae**

Sometimes scales attack the plants, suck the plant sap and thus reduce marketability.

Blue beetle: ***Altica caeruleus*****: Chrysomelidae**

The small grubs mine the cotyledonary leaves and pupate in the leaves.

Tortoise beetle: ***Aspidomorpha exilis*** **:Cassididae**

Eggs are laid singly on ventral surface of leaves. Both grubs and adults damage the leaves. Grubs are flat and spiny and the brownish red oval shining beetle adults feed by scrapping outer tissues of leaves. Pupation takes place on leaf surface. Life cycle is completed in 15 to 30 days. The grubs and adult beetle feed on the leaf by scraping the leaf tissues. As a result, small holes appear on the leaf lamina.

Grasshopper:***Attractomorpha crenulata*** **:Acrididae**

It is a highly polyphagous pest with a very wide range of host plants both cultivated as well as wild. Nymphs and adults nibble leaf lamina causing irregular holes. In case of severe attack, a mild spray with 0.05 % dichlorvos may be resorted to.

Leaf Twisting Weevil: ***Apoderus tranquebaricus*** **:Curculionidae**

The grub of this small weevil cuts across a leaf from margin to midrib near base. Leaf is then folded longitudinally from tip downwards and a compact thimble-shaped structure is formed. This roll gradually starts drying and ultimately falls down along with pupa. Adults are reddish-brown weevils with head drawn anteriorly into a long snout and posteriorly into neck. Adults come out by making a small hole in the dried, rolled mass of leaf. Eggs are oval in shape and yellow in colour. Grubs are apodous and pale-yellow in colour, while pupae are bright yellow.

Aphids: ***Aphids craccivora, Myzus persicae*** **and** ***Lipaphis erisimi*****: Aphididae**

Aphids are sometimes a problem in amaranthus, especially the green coloured varieties, but are easily managed by the application of water spray, sprinkler irrigation, and sometimes, by the use of insecticides with very low residual toxicity applied against other pests.

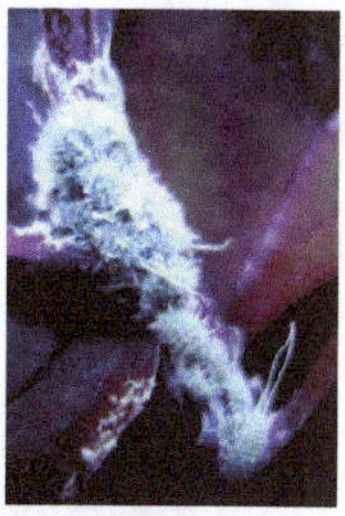

Two tailed Mealy bugs: *Ferrisia virgata* : Pseudococcidae

The adults and nymphs suck vital sap from leaves and shoots.

Thrips: *Euryaplothrips crassus* and *Haplothrips ceylonicus* :Thripidae

The nymphs and adults of the thrips infest inflorescences, sucking the sap and causing flower drop and loss of seed yield, especially in grain amaranth. They also cause crinkling of the leaves and general yield loss in the main crop leaves.

Spider Mites : *Tetranychus* spp. : Tetranychidae

The mites affect the leaves and make them crinkled and cupped. Leaf growth is seriously affected. They can be seen on both surfaces below webbing, with typical symptoms, especially during hot season

Tobacco caterpillar: *Spodoptera litura*: Noctuidae

The caterpillar in the first instar scrapes the leaf epidermis causing papery patches on the lamina. Later it feeds on the leaf voraciously leaving only the main veins and midrib. The grown up caterpillar is stout and smooth with dark colouration and faint longitudinal stripes running along the body.

Gram pod borer: *Helicoverpa armigera*: Noctuidae

The caterpillar is often seen hiding along the inflorescence rachis and feeding on the inflorescence. It is a greenish to brownish smooth caterpillar. The pest is not serious on amaranth, but can reduce yield of grain amaranth.

Semilooper: *Plusia signata*:Noctuidae

The greenish semi looper is found occasionally feeding on the amaranthus leaves

IPM

1. The most important practice is systematic collection and destruction of affected plant parts along with caterpillars.
2. Use light traps to attract and kill adults.
3. Remove all wild amaranthus growth in the vicinity. Give good care for the crop and irrigate well. For many pest problems like thrips and mites, growing green amaranth and red amaranth together will solve a lot of pest and disease problems.
4. In severe attack, harvest the leaves and then spray dichlorvos 0.75 ml/l or other mild insecticides including organic based and natural products of neem. Plan next harvest many days later.

Chapter - 70

BHINDI

Shoot and fruit borer: *Earias vitella (=E.fabia), E. insulana*: Noctuidae

This is a pest on cotton, bhendi, hibiscus, holly hock and other malvaceous vegetables. The damage is due to the larva that bores into tender terminal shoots in the vegetative stage and flower buds; flowers and young fruits in the fruit formation stage. The damaged shoot drops, wither and dry up. The infested fruits have bore holes, are deformed and become unfit for consumption. Bore holes are plugged with excreta.

Adults of *E. insulana* are medium sized, with head and thorax whitish; forewings are pale white with a broad wedge shaped horizontal green patch in the middle. The hind wings are cream white in colour. *E. insulana* adults are smaller than *E. vitella*. Head and thorax are uniformly pea green in colour and forewings uniformly pale yellowish green. They lay about 400 eggs / female. Spherical, light bluish, green, crown shaped, eggs are laid singly on shoot tips, buds, flowers and fruits. The egg period is about 3-5 days. Larva grows for 10-17 days and has six instars. *E. vitella* is brown, with longitudinal white stripes on dorsal side as against cream coloured body with orange dots on prothorax for *E. insulana*. Pupa lasts for 6-10 days and they pupate in a dirty white boat shaped cocoon on the plant itself.

Management

- Collect and destroy infested shoots, buds, flowers and fruits.
- Avoid cultivation of alternate hosts like cotton and hibiscus.
- Release egg parasitoid *T. chilonis* and larval parasitoid *Chelonus blackburnii*.
- Release the lacewing *Chrysoperla carnea*.
- Setup light traps to monitor the moths and their egg laying.
- Setup pheromone traps @ 5/ha.

Spray carbaryl 50% WP 2 g/lit or Malathion one ml/l or Dichlorvos one ml/l or NSKE 50 g/lit of water

Earias vitella adult and larva

E. insulana

Fruit borer: *Helicoverpa armigera* : Noctuidae

This is a medium sized moth yellowish brown in colour with a wing span of about 3 cm and a body length up to 1.4 cm. There is a dark speck and a dark area on the outer margin of each forewing. The forewings also have greyish wavy lines and black spots of different sizes on the upper side. A black kidney shaped mark and a round spot is seen on the underside of the forewing. The hind wings are lighter, whitish, with a broad blackish band along the outer margin.

Young larva feeds on tender foliage and starts infesting the fruits from later instars. They make circular holes and thrust only the head and a part of its body into the fruit and eat the inner contents. The rest of the body is on the outside. A single caterpillar can destroy many fruits

Helicoverpa armigera

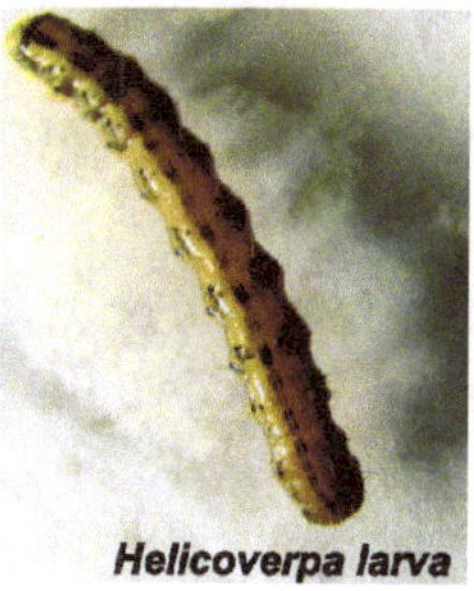

***Helicoverpa* larva**

Spherical yellowish white eggs are laid by the female singly on tender plant parts. Eggs hatch in 2-4 days. Larvae eat for 18-25 days with newly emerged coloured yellowish white and full grown with dark broken grey lines on the side of the body. They have 5 instars. The larvae are cannibalistic in nature and can feed on one another when food is scarce. They are fully fed by 2-4 weeks and are about 3-4 cm long. They are greenish brown with deep yellow stripes, but with many morphotypes based on the host plant and even the surroundings. White and dark grey longitudinal lines and sparse short hairs are seen in many morphs. Pupa: 6-21 days pupation is in earthen cocoon in the soil that takes about 6-21 days. The total life cycle occupies 4-6 weeks with 5-8 generations / year.

Management

- Collect and destroy the infected fruits and grown up larvae.
- Grow simultaneously 40 days old American tall marigold at 1:10 rows to attract *Helicoverpa* adults for egg laying.
- Setup pheromone trap with Helilure at 15/ha and change the lure once in 15 days.

- Conduct six releases of *T. chilonis* @ 50,000/ha per week.
- Release *Chrysoperla carnea* at weekly interval at 50,000 eggs or grubs /ha from 30 days after planting.
- Spray HaNPV or commercial formulations of NPV.
- Spray malathion 50 EC one ml/l or quinalphos 2 ml/l.
- Do not spray insecticides after maturity of fruits.

To encourage activity of parasitoid *Eucelatoria bryani*, *Campoletes*, *Chelonus* etc., use less residual insecticides and prolong the interval between sprays

Leafhopper or green Jassids: *Amrasca biguttula biguttula = A.devastans*: Cicadellidae

This pest has a wide host range and cultivation has to be done with alternate crops as traps crops all around. Spray on the least valued crop grown as the trap crop around the main crop. The host range includes bhendi, brinjal, potato, cotton, sunflower, beans, castor, cucurbits, holly hock and many other malvaceous plants.

The damage is seen as yellowing of leaves followed by crinkling and downward curling leading to bronzing and hopper burn. The adults are greenish leafhoppers. They lay pearl shaped, elongate, yellowish white eggs singly in tissues of main veins on undersurface of leaves.

Eggs hatch in 4-8 days. The nymphs live for 7-10 days. Young nymphs are whitish pale green, wingless, and move diagonally. Adult is wedge shaped, pale green with black dot on posterior part of each forewing. Leaf damage is due to the adult and nymphs sucking the sap from the leaves and also due to the toxaemia produced in response to the saliva injected by the jassids into the leaf and other tissues.

Leaf roller: *Sylepta derogata* : Pyraustidae

This is a pest of cotton, bhendi, hollyhock and other malvaceous plants. The smaller larvae feed on epidermis of ventral surface for two or three days and later roll the leaves and feed within. More than one larva can be seen inside a roll.

Adult moth is with yellowish wings and brown wavy lines. Margins of both wings are bordered with a small layer of short hairs.

Green, round eggs @ about 300/female are laid singly on ventral surface. Eggs hatch in 2-4 days. The larvae take 2-3 weeks for full growth. They are shiny green and transparent with a dark brown head and prothoracic shield. Pupation for 6-12 days Pupates in the roll or rarely descends down and pupates on ground among fallen leaves. Pupae are reddish brown.

Management is usually by regularly collecting and destroying rolled leaves. Rarely insecticides like malathion or dichlorvos may have to be sprayed.

Semilooper :*Anomis flava, Xanthodes (=Acontia) groellsii* : Noctuidae

***Anomis flava* adult**

These are pests of bhendi, cotton and other malvaceous vegetables causing defoliation. These are usually minor pests. The larvae of *A.flava* are greenish with five white longitudinal lines and are brown and medium sized moths with eye spot like pattern while at rest. The larvae of *Acontia groellsii* are green with horseshoe shaped black markings on each segment. The moths are yellowish with v shaped black markings pointing towards the apex while at rest. The larvae of both live for around 25-30 days and pupate in leaf folds and dry leaves for about 8-9 days.

Acontia groellsi

Whitefly: *Bemisia tabaci* : Aleurodidae

The pest is a vector of tomato leaf curl

disease. Nymphs and adults suck cell sap from leaves and cause chlorotic spots resulting in yellowing and drying of leaves. They also secrete honey dew and heavy population will aid development of sooty mould. Severe infestation results in premature defoliation, development of sooty mould and shedding of flowers

If the attack is severe spray dichlorvos one ml/l or malathion 2ml/l. Provide sprinkler irrigation or irrigate the whole plants just as a rain.

Stem weevil: *Pempherulus affinis* :Curculionidae

The grubs bore into stem causing gall-like swellings in the stem, near the base. About 50-120 eggs are laid by each female on the plants. The eggs hatch out in 6-10 days and grubs become grown in 35-40 days. The grubs eat the internal contents and cause gall like swellings on the stem. The adults are dark greyish brown with two white pale cross bands on elytra.

Shoot weevil :*Alcidodes affaber* :Curculionidae

This is another pest which can cause the same symptoms as above. The grubs bore into the stem and petioles causing gall like swellings. The adults feed on the leaves, young buds and also the emerging shoots

Stem maggot :*Melanagromysa obtusa*:Agromysidae

This is basically a pest only in the seedling. The adults lay eggs on the seedlings and the young maggot causes a thickening of the stem around half of the stem, causing a gall like thickening. The growth of the plants is stunted. If the seeds are treated with chlorpyriphos or other seed treatment dusts, this problem and many other pest problems that occur up to around 45 days of the crops can be minimized.

Red cotton bug: *Dysdercus koenigii* F: Pyrrhocoridae

This is not a serious pest in regularly harvested crops. The nymphs and adults such the sap from the young shoots, flower stalks and fruits. If severe, a non residual insecticide like dichlorvos may be sprayed at about 0.5 ml per l. Harvest fruits if any, then spray and allow for two days before the next fruit picking.

Blister beetle: *Mylabris pustulata* : Meloidae

This is one of the common insects that show hyper metamorphosis. There are different shapes for the different instars. The adult feeds voraciously on buds and flowers. A single beetle can destroy as many as 20-30 flowers/day. The eggs are laid by female beetle in clusters of 60-80 eggs at 2-3 cm depth in soil. Eggs are light yellowish in colour and cylindrical in shape. Incubation period is about three weeks. Young grubs are white in colour. The first instar is a triungulin. Then the first grub phase comes in. In the sixth or seventh instar, the coarctate phase is seen and /or the second grub phase. Then it pupates inside the soil tunnel and adult emerges.It is more important in pulses than in bhendi.

Dusky cotton bug – *Oxycarenus hyalinipennis*: Lygaeidae,

The nymphs and adults of this small bug are sometimes seen on bhendi. This is primarily a pest of cotton. On bhendi, the pest sucks the sap from the young shoots, petioles and then the fruits. Usually managed by monitoring, hand collection and later on by the application of neem formulations. The seeds will also be discoloured and shrunken.

Cotton Aphid :*Aphis gossypii* : Aphididae

Adults and nymphs suck the sap from emerging shoots, leaves and fruits. This is a minor pest.

Red spider mite: *Tetranychus telarius* : Tetranychidae

This is a common summer pest many plants like bhendi, cotton, tomato, brinjal, castor, gourds, tea, citrus, grapes, rose, jasmine and marigold. The Damage is caused by both the nymphs and adults that feed on the ventral leaf surface, under a protective cover of fine silken webs, as a result of which innumerable yellow spots appear on the dorsal side of leaves. Affected leaves gradually start curling, finally get wrinkled and crumpled. Mites penetrate the plant tissues by surface scratching by the mouth parts and suck the exuding sap. The leaves lose their turgor and the photosynthetic efficiency is lost and later dries up fully. Adults are reddish or yellowish brown with blackish legs. Eggs hatch in 3-5 days and they feed for 6-10 days before they become the adults and start depositing fresh eggs under a fine web.

Management: Application of wettable sulphur 80WP @ 2g/l or Dicofol or Dimethoate 30 EC 1.5ml/l or Phosalone 2ml/l on the under surface of the leaves. Provide copious irrigation on the under surface of the leaves.

Other pests

- **Leaf weevil: *Myllocerus* sp. : curculionidae.**
- **Leaf miner :*Trachys herilla* : F. Buprestidae.**
- **Grasshoppers :*Poekilocerus pictus*, *Oxya japonica*: Acrididae.**
- **Dusky cotton bug: *Oxycarenus hyalinipennis* : F. Lygaeidae.**
- **Mealy bug :*Ferrisia virgata* : Pseudococcidae.**
- **Scales :*Saissetia coffeae*, *Parasaissetia nigra* : Coccidae.**
- **Chafer beetle :*Oxycetonia versicolor* : F. Cetonidae.**

Chapter - 71

BRINJAL

Shoot and fruit borer: *Leucinodes orbonalis*: Pyraustidae

The pest is active throughout year, but more during monsoon period. Heavy losses up to 60% can occur.

The adult is a greyish brown medium sized moth with white and hyaline wings, having triangular brown and red markings on forewing. A crescent shaped mark facing outside at the apical margin is typical for the species. About 150-250 creamy white eggs are laid singly on leaves, tender shoots, flowers and developing fruits. Egg period is 3-5 days and the hatched larva completes its life in 15 days with 5 instars. They are stout, pink coloured with sparsely distributed hairs on warts on the body and with brownish head. Pupation is in a boat shaped tough greyish cocoon for 6-8 days on the plant itself. Total life is completed in about 17-50 days.

On plants in the pre flowering stage, the larvae bore into tender shoots with the holes plugged with excreta. This causes withering of terminal shoots /

dead hearts. They also bore into petioles of leaves, flower buds and developing buds causing withering of leaves and shoots. If the attack is in the later stages, the attack causes shedding of buds and make fruits unfit for consumption. Attacked fruits are with bore holes plugged with excreta. Fruits become malformed also. Yield loss from 60-80 % can happen due to the pest.

Management

- Observe for the presence of the pest everyday. Collect and destroy the damaged tender shoots, fallen fruits and fruits with bore roles at least once a week. An IPM module also includes use of pheromone traps @40/ha(roughly 2/10 cents) to prevent population buildup, along with application of NSKE (4%)spray once in two weeks.
- Avoid ratooning and continuous cropping of brinjal crop. Burn old crop remnants and plant some other vegetables.
- Grow resistance varieties, best adapted for each locality and with consumer preference.
- Use light traps @ 1/ha to attract and kill insects.
- Spray formulations of *B. thuringiensis* var *kurstaki.*
- When insecticide is needed, malathion 2ml/l or dichlorvos one ml/l or Neem oil 2 ml / l can be sprayed.
- Avoid using highly residual insecticides at the time of fruit maturation and harvest.

Epilachna beetle/Hadda beetle: *Henosepilachna viginti octopunctata*: Coccinellidae

Adults of these species are distinguishable with the number of spots on the elytra***. *E. vigintioctopunctata*: 14 spots on each elytra, deep red. Adult of *E. dodecastigma*: Copper-coloured, 6 spots

***According to taxonomists, the spots are not an identifying character. The species feeding on brinjal, is *H. viginti octopunctata,* according to them

Both adult and grubs scrape the lower epidermis of leaves leaving behind riddled areas with window veined papery areas. In severe infestation all leaves may be eaten off leaving only the veins intact (Skeletonization), with the leaves or whole plants withering.

Adults lay 120-460 eggs/female. They are cigar shaped, yellow and are laid in clusters on lower leaf surface. The eggs hatch in 2-4 days The grub period is 10-35 days. They are yellowish bearing six rows of longitudinal spines with terminal branching. Pupa occupies 5-6 days and is seen on the stem or leaves.

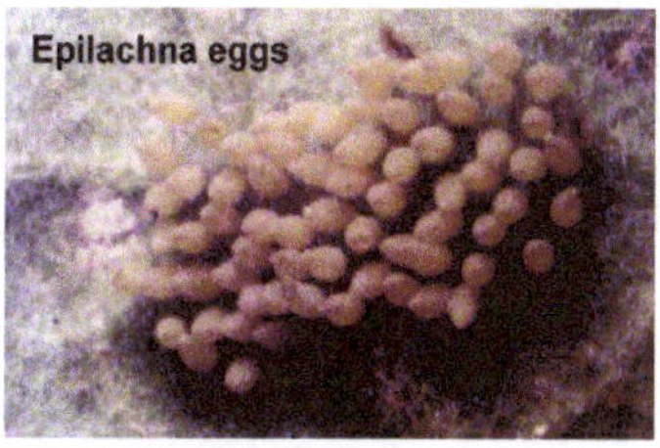

Total life period occupies 20-35 days with 7 -10 generations / year.

Management

- Systematically collect and destroy leaves with egg masses, grubs and pupae and also adult beetles
- Spray Dichlorvos one ml/l or malathion 50 EC 2 ml /l when populations explode.
- Diflubenzuron or Buprofezin also works well on the beetles

Stem borer: *Euzophera perticella*: Pyralidae

Adults are Greyish brown, forewings with transverse blackish wavy bands of different thickness and white hind wings. Cream coloured, scale-like eggs are laid singly or in batches on young leaves, petioles and branches and emerge in 10 days. The pale grey larvae

bore into petioles or main stem of young and old plants and move downwards. Top shoots of young plants wilt and wither. Older plants become stunted. Fruit bearing capacity is adversely affected. There is a distinct thickening at the entry point. The larvae live for about 26-35 days. Fully grown larva is creamy white with few bristle-like hairs, around 20 mm in size. They pupate within a cocoon inside the larval tunnel for 10-16 days. Take remedial measures as in the case of the shoot and fruit borer.

Ash weevils: *Myllocerus subfasciatus*, *Myllocerus laetivirens*, *M. undecimpustulatus*, *M. discolor*, *M. Viridanus*, *Amylyrrhinus poricollis* :curculionidae

M. subfasciatus is brownish, M. *discolor* is brown with white spots while *M. viridanus* is a light green weevil.

The adults and grubs cause damage. Young larvae (grubs) feed on the roots of plants. The adult weevils cut holes on the leaf or nibble the margins of leaves

The grubs feed on the roots of plants and cause wilting and drying of plants. They also girdle the stem. But these weevils are generally of minor importance, due to low populations of both the grubs and adults, and do not require specific chemical management.

Biology: The adult weevils lay about 300-500 eggs in the soil over a period of three months. The grubs hatch out in a week and start feeding on the roots of the hosts. The grub period is around 60-75 days. The full grown grubs pupate in earthen cells for 10-12 days. The emerging adults take to the aerial portions and start feeding on the leaves of the hosts. They are polyphagous and can feed on a variety of crop plants and other vegetation.

Management

In cases of considerable damage to the roots in a plantation, the root zone can be drenched with 0.01 % Chlorpyriphos or rake up a dust formulation of Chlorpyriphos or any other dust formulation in the soil. The adults can be warded off or killed by application of neem based insecticides or contact insecticides like Malathion 0.05 % or Quinalphos 0.05% or synthetic pyrethroids at 0.01-0.05 % depending on the chemical nature. Unless they cause significant damage, chemicals should not be applied.

Brinjal mealy bug: *Coccidohytstrix insolitus*: Pseudococcidae

These are white mealy bugs that infest the leaves and tender shoots in very large numbers on the grown up plants. They are attended by ants. The leaves turn yellow and unproductive. Destroy affected leaves as soon as they are noticed. In severe infestation, spray Quinalphos 25 EC 2ml/l after harvesting all ripe and semi ripe fruits. The next harvest should be prolonged.

Brinjal Lacewing bugs: *Urentius hystricellus*: Tingidae

These are soft insects and the adults and nymphs are brownish and spiny, with lace like wings, about 3.5 cm in length. The eggs are inserted into the leaves, and hatch in around a week. Nymphs and adults desap the plants, adults on the upper side and nymphs on the lower side of leaves. The leaves dry up. If needed, give a spray of Quinalphos 25 EC 2ml/l.

Leaf webber: *Psara bipunctalis*: Pyralidae

This is a small moth, straw coloured, with wavy lines and black dots on the wings. The adult has a wing span of around 2 cm. The caterpillar is greenish and attains a length of 2.5 cm when full grown. The larvae of the leaf webber feed from within the webbed leaf. The only remnants are the webbings, excreta and the leaf veins. The pupa is seen inside the webbing. Collect and destroy the webbings as soon as they are seen.

Sphingid caterpillar: *Acherontia styx*: Sphingidae

The adults are big moths and the larvae are also giant caterpillars. Adults have a wing span of about 11 cm and can be 5 cm long. The fore wings are grey brown or blackish with wavy transverse markings. The hind wings are yellow and have dark patches along the border. The eggs are laid singly on the leaves. The eggs hatch in 2-5 days. The larvae feed voraciously on the leaves for about 2-3 weeks and become stout and sturdy, are

greenish and with dark green oblique lines on its sides. There is a curved horn on the dorsal side of the tail end of the larvae. Pupation is in the soil, taking 2 weeks for the adult to emerge. To prevent extensive defoliation, the larvae have to be hand collected and destroyed.

Leaf roller : *Antoba(Eublemma) olivacea*: Noctuidae

This is a medium sized moth with olive green colouration. The larvae are hairy and purple brown with yellow spots. They fold the tender leaves and feed from within. The pupa is also inside. Collect and destroy the larvae in the leaf folds as and when they are formed.

Antoba olivacea

Hairy caterpillar – *Selepa celtis* and *Selepa docilis*: Noctuidae

The adult is a small grayish brown moth with white hind wings. Eggs are laid in groups in the leaves. The hairy yellow larvae feed gregariously in the upper leaf surface in the beginning. Later they disperse and eat the leaf fully. Pupation is on the plant as a cocoon. Life cycle takes about a month.

Selepa docilis

Midge: *Asphondylia* sp.:Cecidomyiidae

The pest attacks flower buds or young fruits in the larval stage as maggots, causing drying of flowers or malformation of fruits. Loss is severe during cooler months. Up to 70% flowers are damaged and fail to set fruits.

Bud- worm/ Potato tuber moth: *Phthorimoea operculella*: Gelechidae

Small pink larvae of these feed inside flower buds and reduce fruit formation.

Mealy bug: *Coccidohystrix (=Centrococcus) insolitus*: Pseudococcidae.

The white mealy bugs infest the leaves and tender shoots in hundres during the later stages of the crop. They are also attended to by small ants. The leaves become unhealthy, yellowish and dry up. Collect and destroy the leaves as and when they are being attacked.

Red spider mite: *Tetranychus telarius (T. urticae)*: Tetranychidae

Damage

Nymphs and adults feed on ventral leaf surface, under protective cover of fine silken webs, and as a result of their feeding, numerous yellow spots appear on dorsal side of leaves. The infested leaves become discoloured, gradually start curling, finally wrinkled and crumpled. The yield is reduced drastically.

Eggs are laid on the leaves, and egg period is 3-5 days. The nymphs live for about 6-10 days, and the ovate, reddish brown adults, nymphs and eggs are all seen in the leaf on the undersurface beneath a webbing.

Spray wettable sulphur 80 WP 2 g/l (or) Dicofol 3 ml/l or thiomethoxam, when the population is heavy. Apply copious irrigation to the leaves.

Other pests

Aphid: *Aphis gossypii, Myzus persicae*: Aphididae

Colonies of aphids are found on tender leaves, flower buds, etc. Generally, aphid population can increase due to indiscriminate use of insecticides.

Red bug: *Aspongopus janus*: Pentatomidae

Medium sized reddish flat bugs are found sucking sap on tender parts.

Leafhopper:*Amrasca devastans (=A. biguttula biguttula)*: Cicadellidae

The nymphs and adults suck the sap from the underside of leaves. The Infested leaves curl upward in the margins.

Thrips:Onion thrips : *Thrips tabaci;* Melon thrips: *Thrips palmi*: Thripidae

Both nymphs and adults lacerate tender fruits and leaves. However, they usually congregate on lower the surface of older leaves.

Termites: *Microtermes* sp., *Trinervitermes biformis*: Termitidae

These are usual in dry sandy soil. The termites eat the roots and cause sudden wilting of seedling.

Chapter - 72

CRUCIFEROUS VEGETABLES (CABBAGE, CAULIFLOWER AND OTHERS

Diamond back moth: *Plutella xylostella*: Plutellidae

The adults of this moth are small, greyish brown having pale whitish narrow wings and with inner margins yellow. There are three pale whitish triangular markings on hind margins of each forewing. At rest, when both forewings come together, they form a dorsal median patch of three diamond shaped yellowish white spots that give them their name diamond back moth. Hind wings have a fringe of long hairs.

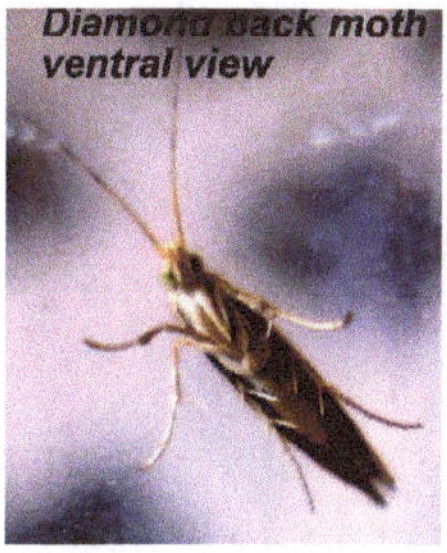

First instar larvae mine epidermal surface of leaves producing typical white patches. Larvae, from second instar onwards feed externally making holes on the leaves and also contaminate them with excreta. Heavy infestations cause shot hole like appearance and retardation of the growth of cabbage and cauliflower. Eggs are laid on the undersurface of leaves along the veins. Up to 60 eggs are laid per female. Larva is 8-12 mm long, pale yellowish green in color, pointed at both the ends with fine erect black hairs scattered all over the body. Pupation is on the underside of the leaves and one life cycle per month is possible.

IPM

- There are many larval and pupal parasitoids for the pest in nature. Initial attack itself has to be monitored and managed in every crop. Grow mustard all around as trap crop and use pheromone traps for monitoring and management. When adult population goes up substantially, spray the trap crop with malathion 2ml/l or dichlorvos one ml/ l or quinalphos 2ml/l or fipronyl 2ml /l. Mustard as trap crop can be planted in advance. Raise 2 rows of mustard for every 25 rows of cabbage. Sow first mustard crop 15 days prior to cabbage planting or plant 20 days old mustard seedling at the time of planting the main field. Usually,30-35 day old cabbage seedlings are used in the main field for transplanting.
- Apply formulations of *Bacillus thuringiensis*.
- On the main crop also , malathion 2ml/l or dichlorvos one ml/ l or quinalphos 2ml/l or fipronyl 2ml /l or cartap hydrochloride 1 g/l or carbosulfan 2 ml or quinalphos 2ml/l can be sprayed at 30-36 day age to prevent loss at head formation .
- The pest is regulated by two releases of larval parasitoids viz., *Cotesia plutellae* in plains and *Diadegma* spp. in hills commencing from 20 days after planting to reduce the damage.

The cruciferous Leaf webber: *Crocidolomia binotalis*: Pyraustidae

This is a serious pest of crucifers including cabbage, radish, cauliflower and mustard. The Young larva feeds gregariously on leaves, with silken strands and later webs together the leaves and feeds from the undersurface, completely skeletonising them. Due to gregarious feeding, rotting of cabbage heads and cauliflower curds are common. Eggs are laid in masses of 40-100, on the lower side of leaves, and develop fully within 24-27 days. They are pale green, with red head, brown longitudinal stripes and rows of tubercles with short hairs. Adults are small pale brownish, with distinct wavy lines and prominent wavy spots on forewing. Hind wings are semi-hyaline.

Management

In severe infestations, spray phosalone 35 EC 2 ml/l, Carbaryl 4gm /l ,fenvalerate 20 EC 0.5 ml/l, cypermethrin 10 EC 0.5 ml/l, deltamethrin 28 EC 0.5 m/l, or cartap hydrochloride 50 WSP 1 g/l. Sprays against DBM would suppress leaf webber also.

Cabbage borer: *Hellula undalis*: Pyraustidae

This is only a sporadic pest, but occasionally becomes serious and aborts head formation. The caterpillars in their first two instars are leaf miners alongside the veins, causing a papery white leaf with remains of excreta on it. Later on they feed on leaves or shoot and finally bore into the stems. They prevent head initiation causing multiple shoots or heads.

The moth has greyish brown forewings with greyish white wavy markings and the hind wings are powdery pale. There is a pale apical spot and pale dark area on the forewings. The larvae are pale whitish-brown with 4-5 purplish brown longitudinal stripes. Eggs are laid in clusters up to 20 eggs on the undersurface of leaves, or sometimes on tender shoots or other plant parts. They get fully grown in about 6-7days. The head and prothracic shield are blackish brown with dark spots and the body has sparse hairs on minute tubercles.

Bracon hebetor parasitizes the pest in nature, and heavy incidences can be controlled by sprays of malathion 2ml/l or dichlorvos one ml/ l or quinalphos 2ml/l or fipronyl 2ml /l or cartap hydrochloride 1 g/l or carbosulfan 2 ml/l or quinalphos 2ml/l.

Cabbage aphid: *Brevicoryne brassicae*: *Aphididae; Mustard aphid*: *Lipaphis erysimi*: Aphididae

Colonies of aphid are found on tender shoots and suck sap from plant tissues. In case of severe infestation plants may completely dry up and die off. Feeding damage from large numbers of aphids can kill seedlings and young transplants. On larger plants, feeding by aphids results in curling and yellowing of leaves, stunted Plant growth, and deformed heads. The aphids cast off their exuvium or skin periodically and this will be seen in the colony itself, amidst aphids as

white cast skin .The nymphs and adults of this aphid are yellowish green with wavy white filamentous markings over the body. For management, it is always good to use yellow sticky traps in the field. They attract and stick the aphids. Usually about 4 -6 traps can be kept, but many pieces of yellow sheets smeared with oil serve well to reduce aphids and also to monitor them. If the infestation is very severe, spray any one of the following:

- DDVP or dichlorvos 1ml/l
- Malathion 1ml/l
- Neem oil 800 ml/acre

Gram- caterpillar: *Helicoverpa armigera*: Noctuidae

This is one of the most ubiquitous pests throughout the world. The initial larvae may scrape the leaves, but the later instars cut big holes plugged with excreta on formed heads. The pest is polyphagous on many crops. For damage and management, see cotton, pulses or tomato.

Green semi looper: *Trichoplusia ni* : Noctuidae

The adult is a stout moth, medium sized, with grey head and thorax, grey wavy forewings and with an abdominal tuft of reddish green hairs. The larvae are slender and long in the anterior region, with a broad lateral stripe and with light wavy markings on the abdomen. They eat the leaves and render it skeletonised.

Cabbage butterfly: *Pieris brassicae, P. rapae*: Pieridae

These are white butterflies and the larvae of these are velvety, bluish green, and with black darts. There are yellow dorsal and lateral stripes and also has white hairs on the larval body. The adults of these have black bordered white forewings with typical colour patterns

Cutworm: ***Agrotis ipsilon*** **:Noctuidae**

The Adult has pale brown forewings with a dark purplish brown tinge on the costal margin. The hyaline hind wings have a tinge of brown. The male has a distinct bipectinate antenna.

Caterpillars are nocturnal in habit and hide during day in cracks and crevices in soil or under debris around plants. At night they come out, cut seedlings near ground level and eat tender parts. Full-grown caterpillars enter soil and pupate in earthen cocoons. Egg, caterpillar and pupal stages last for 2 to 3, 10 to 30 and 8-10 days, respectively. Total life cycle is 30 to 68 days.

Other cutworms: ***A. segetum*****:** ***Xestia C-nigrum and Peridroma saucia*** **:Noctuidae**

These are polyphagous pests with a wide range of host plants including cole crops. Other hosts are potato, barley, beet-root, gram, okra, linseed, lucerne, millets, oats, peas, poppy, pulses, tobacco and wheat. They function as cutworms as described above.

Leaf eating caterpillar: ***Spodoptera litura*****: Noctuidae**

The adult is a stout greyish brown moth 3.8x1.8 with wavy white markings and criss-cross lines on the forewings and white hind wings having a brown line at the peripheral fringe. Young larva scrape leaves on ventral surface, and growing larvae defoliate crops and also feed on young fruits.

Eggs are laid in masses covered with white hairs on the surface of affected leaves. Eggs hatch in 4-5 days. Larva eats for 14-20 days. They are pale brown with greenish to violet tinge and with stout whitish or creamy bands running the length of the body. Yellow and purplish spots are found in sub marginal areas. Pupa is seen in the soil for 7-15 days.

Management

- Plough the soil to expose and kill the pupae.
- Grow castor along borders or irrigation channels as indicator or trap crop.
- Flood the field to drive out hibernating larvae.
- Set up light trap @5-6/ha for monitoring and kill. Also use pheromone trap at 15/ha with pheroclin SL lure.
- Destroy egg masses and grown up larvae.
- Spray SlNPV at 250 LE along with teepol 1ml/l in the evening.

Spray Quinalphos 25 EC 2ml/l or Malathion 2ml/l or Dichlorvos 76 WSC 1 ml/l or Endosulfan35EC 3ml/l or NSKE 5% during evening hrs.

Mustard sawfly: *Athalia lugens proxima*: Tenthridinidae

This is one of the rare species in hymenoptera that are crop pests. The pest is so denoted because of the saw like ovipositor. The adults are orange and black with translucent smoky wings and black veins. The head is blackish and the thorax is orange coloured. Larva is greenish black with wrinkled body and eight pairs of pro-legs. On slightest touch the larva falls to round and feigns death. The larvae attack the cole crops by nibbling the margins of the tender leaves and later on biting holes on it, rendering it as shot holed leaves. A pesticide spray is needed in case of severe attack.

Radish flea beetles:*Phyllotreta downsei, P.cruciferae, Chaetonema basalis* :Alticidae

These are small beetles that are usually night feeders or dusk feeders. The adults are the damaging stage and nibble the tender leaves to produce holes on the leaves. These are about 1 - 2 mm long, with bluish black or shiny black colour. They can easily be managed by applying a less residual pesticide dust formulation.

Tussock caterpillar: *Orgyia postica*: Lymantridae

The caterpillars of this pest are found feeding on the leaves of cabbage and cauliflower in the plains of Kerala where the farmers have started growing the varieties suited for mild cool tropical winter.

Painted bug: *Bagrada hilaris (cruciferarum)*: Pentatomidae

The nymphs and adults of this bug suck the sap from the leaves, shoots and other parts of the crop. They emit the typical bug odour when disturbed.

Other minor pests are the **Pea leaf miner**,***Chromatomyia (phytomyza) harticola***(Agromyzidae) and **The geometrid**:***Xanthorrhoea saturata***:**Geometridae**

Chapter - 73

CHILLI AND CAPSICUM

Chilli thrips: *Scirtothrips dorsalis, Thrips palmi*: Thripidae

The adults of this insect are small brownish black with fringed wings. The nymphs are slender and yellowish straw coloured, later acquiring more brownish hue. They lacerate the tissues and feed on the exuding sap. Leaves become crinkled, curled upward and shed. The attack can be confused with chilli mosaic symptom. In capsicum and chilli, they also feed beneath calyx of flowers and fruits causing flower/fruit drop and reduction in yield and market value of capsicum. Buds become brittle and drop down. Plants get stunted and bronzed. The fruits if at all formed are crinkled and small.

These can reproduce sexually as well as parthenogenetically. Female thrips insert the eggs into the veins and a female lays upto 40-48 eggs. Lifecycle occupies 10-20 days.

Management

- These are summer pests. Irrigate the crop canopy to reduce the temperature. If needed, spray crop with malathion one ml/l or dimethoate 30 EC 1.5 ml/l. Root dip the seedlings in Imidachloprid20 EC-3-5 ml/ 10 L, Thiacloprid one ml/L overnight before transplanting

Spiralling whitefly/cottony whitefly: *Aleurodicus dispersus*: Aleyrodidae

This pest originated from the tropical countries in Central and South America. More recently, it has been reported from Asia and Africa. In India, Andhra Pradesh, Karnataka, Kerala, Maharashtra and Tamil Nadu faces its problem. *A. dispersus* is a highly polyphagous species. Its wide host range includes many vegetables including *Capsicum*, ornamental and fruit crops, as well as numerous trees and shrubs, *Citrus*, *Cocos nucifera* (coconut), *Euphorbia pulcherrima* (poinsettia), *Glycine max* (soybean), *Hibiscus*, *Lycopersicon esculentum* (tomato), *Mangifera indica* (mango), *Musa* (banana), *Persea americana* (avocado), *Prunus* spp., *Psidium guajava* (guava)and *Solanum melongena* (brinjal) etc.

Egg spiral of spiralling whitefly

Immature and adult stages of *A. dispersus* cause direct feeding damage by sucking plant sap, which can cause premature leaf fall. Indirect damage is due to the heavy production of honeydew and white, waxy material produced by the insect. Sooty

Pupa adults and nymphs of spiralling whitefly in waxy coating

mould develops on honeydew and decreases photosynthetic activity. Plants are also disfigured and may be unmarketable. In places where it occurs, *A. dispersus* is generally considered as a serious pest, causing crop losses. Virus transmission is apparently not known.

Natural dispersal can be through flying adults. Over long distances, the pest has already showed its potential for spread, being introduced into many different parts of the world. Movements of infested plants or fruits can ensure long distance dissemination.

Biology: The adults lay hundreds of eggs within the colony on the undersurface of the leaves. The eggs are laid in a characteristic spiral shape and this gives the species its name as the spiralling whitefly. These emerge within a few days and the young ones emerge out. These are called as crawlers and are the moving stage of the pest. They move out, by actively crawling out and find out a suitable site for feeding. Once settled, they feed, and go for the first moult. They do not normally move out very farther in their life time. The nymphs are all coated with a white waxy layer and have filamentous extrusions from the body. It is thus also appropriate to call them the cottony whitefly or as mealy whitefly. The next is the second instar nymph, and two more moults occur, the last one transforming the nymph into a so called puparia. The puparium in close up, is a very decorative piece, resembling a pure white decorated icing cake, with silver threads tastefully arranged and protruding out. All the nymphal stages are phloem feeders. The nymphs secrete a waxy, filamentous, curly material over their body .The adult fly looks like a miniature wax coated moth, and can mate and later fly off to other plants and get dispersed by air currents. They can lay eggs on fresh plants. The natural control of the pest is by predatory lady bird beetles like *Cryptolaemus montrouzieri* or lacewings or by parasitoids like *Encarsia* spp. Chemical control is difficult, but contact insecticides, like Quinalphos 2ml/l or malathion one ml/l , with addition of soap solution @ 5ml /l of spray solution or neonicotinoids like imidachloprid @3ml/10 l will give reasonable control. Reducing the temperature of the microclimate by spraying water on the undersurface of the crops, especially on vegetable crops, reduces the pest. This pest is most abundant on the undersurface of guava and Teak leaves.

Green peach aphid: *Myzus persicae*: Aphididae

These aphids congregate on the undersurface of leaves and suck the sap. All life stages of nymphs and adults desap from the leaves and cause curling and crinkling of leaves and young shoots. The leaves also show a coating of honeydew and subsequently, sooty mould. Plants remain stunted. Adult is mostly yellow in colour.

Myzus persicae

Management

Mechanical removal of the first colonies and dislodging the aphids by a jet of water are the preliminary steps .In severe cases, apply Methyl demeton 25 EC 1ml/l ,dimethoate, 30 EC 1 ml/ l or neem based formulations 2 ml/l.

Tobacco caterpillar: *Spodoptera litura*: Noctuidae

The adult is a stout greyish brown moth 3.8x1.8 with wavy white markings and criss-cross lines on the forewings and white hind wings having a brown patch along its margin. Young larva scrape leaves on ventral surface, and growing larvae defoliate crops and also feeds on young fruits.

Spodoptera litura on capsicum

Eggs are laid in masses covered with white hairs on the surface of affected leaves. Eggs hatch in 4-5 days. Larva eats for 14-20 days. They are pale brown with greenish to violet tinge and with stout whitish or creamy bands running the length of the body. Yellow and purplish spots are seen in sub marginal areas. Pupa is seen in the soil for 7-15 days.

- Set up light trap at 4-5/ha or pheromone trap at 15/ha with pheroclin SL lure.
- Destroy egg masses and grown up larvae.
- Spray SlNPV at 250 LE along with teepol 1ml/lit in evening hrs.

Spray Quinalphos 25 EC 2ml/l or Malathion 2ml/l or Dichlorvos 76 WSC 1 ml/l or Endosulfan35EC 3ml/l or NSKE 5%. Spray the insecticide during evening hrs.

Greasy/black cut worm: *Agrotis ipsilon*: Noctuidae

The pest is so named due to its habit of coming out during night and cutting the seedlings at ground level and eating the tender leaves. Adult has a pale brown forewing with dark purplish brown area along costal margin and base of hind wing white with brown tinge. Male has a bipectinate antenna. Larvae are black with pale mid dorsal stripes and whitish underbelly. The head is pale brown.

- In endemic areas, plough the soil during summer months to expose larvae and pupae to predators .Collect adults in light traps.Use pheromone traps @ 12/ ha to attract male moths.
- A poison bait can also be prepared with mistened rice bran, molasses and an insecticide like malathion or carbaryl(10:2:1). The moist bait can be made into globules and kept in and around the field for the larvae to eat and die
- Apply insecticides like neem products, NSKE, malathion or dichlorvos in the evening hours at the base of chilli and on the foliage.

Fruit borer: *Helicoverpa armigera*: Noctuidae

This is a medium sized moth yellowish brown in colour with a wing span of about 3 cm and a body length up to 1.4 cm. There is a dark speck and a dark area on the outer margin of each forewing. The forewings also have greyish wavy lines and black spots of different sizes on the upper side. A black kidney shaped mark and a round spot is seen on the underside of the forewing. The hind wings are lighter, whitish, with a broad blackish band along the outer margin.

Young larva feeds on tender foliage and starts infesting the fruits from later instars. They make circular holes and thrust only the head and a part of its body into the fruit and eat the inner contents. The rest of the body is on the outside. A single caterpillar can destroy many fruits

Spherical yellowish white eggs are laid by the female singly on tender plant parts. Eggs hatch in 2-4 days. Larvae eat for 18-25 days with newly emerged coloured yellowish white and full grown with dark broken grey lines on the side of the body. They have 5 instars. The larvae are cannibalistic in nature and can feed on one another when food is scarce. They are fully fed by 2-4 weeks and are about 3-4 cm long. They are greenish brown with deep yellow stripes, but with many morphotypes based on the host plant and even the surroundings. White and dark grey longitudinal lines and sparse short hairs are seen in many morphs. Pupa: 6-21 days pupates in earthen cocoon in soil Total: 4-6 weeks with 5-8 generations / year.

Management

- Collect and destroy the infected fruits and grown up larvae.
- Grow simultaneously 40 days old American tall marigold at 1:10 rows to attract *Helicoverpa* adults for egg laying.
- Setup pheromone trap with Helilure at 15/ha and change the lure once in 15 days.
- Six releases of *T. chilonis* @ 50,000/ha per week coinciding with flowering time.
- Release *Chrysoperla carnea* at weekly interval at 50,000 eggs or grubs / ha from 30 days after planting.
- Spray HaNPV or commercial formulations of NPV.

- Spray malathion 50 EC one ml/l or quinalphos 2 ml/l.
- Do not spray insecticides after maturity of fruits.

To encourage activity of parasitoid *Eucelatoria bryani*, *Campoletes*, *Chelonus* etc., use less residual insecticides and prolong the interval between sprays

White grub:*Holotrichia consanguienea*:*H.reynaudi* ; *Anomala bengalensis*: Scarabaeidae

Eggs are laid in soil during pre-monsoon showers and grubs feed on roots. Adults feed on leaves.

Sun hemp Hairy caterpillar: *Utetheisa pulchella*: Arctiidae

The adult is a small beautiful tiger moth, attractive in the sitting posture, and also when it flies. It has white wings with red and black spots on the fore wings and black marginal designer wavy band on the hyaline hind wings. The caterpillar has a body with red, black and white spots, is thinly hairy and has yellow lines on the dorsum/ dorsolaterally. These caterpillars eat the surface layer of the fruits.

Broad mite/ yellow mite: *Polyphagotarsonemus latus*: Tarsonemidae

P. latus has a world-wide distribution, and is known by a number of common names. In India it is called the "yellow tea mite,". Female mites are about 0.2 mm long and oval in outline. Their bodies are swollen in profile and a light yellow to amber or green in color with an indistinct, light, median stripe that forks near the back end of the body. Males are similar in color but lack the stripe. The two hind legs of the adult females are reduced to whip-like appendages. The male is smaller (0.11mm) and faster moving than the female. The male's enlarged hind legs are used to pick up the female nymph and place her at right angles to the male's body for later mating

The eggs are colorless, translucent and elliptical in

shape. Adult females lay 30 to 76 eggs (averaging five per day) on the undersides of leaf surface and in the depressions of small fruits. Nymphs have only three pairs of legs and are whitish. After one day, the larva becomes a quiescent nymph that is clear and pointed at both ends. The nymphal stage lasts about a day. The adults and nymphs cause terminal leaves and flower buds to become malformed. The mite's toxic saliva causes twisted, hardened and distorted growth in the terminal shoots. They congregate on the young shoots and fruits. Population is more on the newest leaves and small fruit. Leaves turn downward and turn coppery or purplish. Internodes shorten, buds and developing fruits drop and yield is affected. Insecticidal oils or soaps are effective and less toxic to the environment. If the pest is severe,spray Dicofol 18.5 EC @ 2.5 ml/ l or phosalone 35 EC 2 ml / l or wettable sulphur 50WP @ 6 g m / l of water

- For greenhouse control, biological control agents are available, including several species of predatory mites like *Amblyseius ovalis.*

Mites: *Tarsonemus translucens*: Tarsonemidae,*Tetranychus cinnabarinus*: Tetranychidae

These are small mites, and suck the sap from the underside of leaves. The leaves curl downwards. Very serious in chilli and can cause up to 80% loss in chilli if left uncared for.

Other pests

Fruit fly: *Bactrocera dorsalis* , *B.Cucurbitae* :Tephritidae

Gall fly: *Asphondylia capsisi*: Cecidomyidae

Yellow aphid :*Aphis gossypii*: Aphididae

Cotton whitefly: *Bemisia tabaci*: Aleyrodidae

Fruit bug: *Lygaeus hospes*: Lygaiedae

Two tailed mealy bug: *Ferrisia virgata*: Pseudococcidae

Hard scale :*Aspidiotus destructor*: Diaspididae

Pepper scale: *Lepidosaphes piperis*: Diaspididae

Fruit sucking moths :*Othreis spp*:. Noctuidae.

Chapter - 74

COWPEA AND OTHER PULSES

The cowpea aphid: *Aphis craccivora*: Aphididae

Hosts

Groundnut, red gram, peas, beans, safflower, lab-lab niger

Symptoms of damage

Both nymphs and adults cause the damage by sucking the plant sap from shoot tips, pods and leaves. Infested pods become malformed and withered. Severe infestation may result in complete drying of affected pods. They also act as vector of virus diseases in pea.

Bionomics

It is a greenish black coloured aphid. The total life cycle from first instar nymph to adult occupies an average of 8-10days. It reproduces parthenogenetically and viviparously. The female may produce 10-30 young ones in a life span of 10-12 days. The nymphs transform into adult in 5-8 days after passing through four moults.

Management

1. Use entomopathogenic fungus *Fusarium pallidoroseum* to cause

epizootics in aphids in the cowpea field.

2. Spraying of infested crop with 0.05% Malathion or dichlorvos with a spray fluid of 500 litre water per ha effectively control aphids. This pest has a quick multiplication rate and hence the insecticidal treatment has to be repeated as soon as aphid population is found to have built up again.

The papaya Mealy bug :*Paracoccus marginatus* :Pseudococcidae

The papaya Mealy bug *Paracoccus marginatus* has created a serious problem for many crops and non crops. The pest has a cottony white powdery (mealy) coating over its body. Some farmers and common people have even described it as a fungal infection because of the white powdery mass over the insect and also because the insect does not move at all, once it develops into the adult stage. There is also the white powder that might fly off, so that the coating is thought of as fungal spores or mycelium.

The papaya mealy bug is a native of Mexico and/or Central America. It has never been a pest there, due to the presence of a large number of natural enemies in their homeland. The first records were from Mexico in 1955. It was then described in 1992 from Belize, Costa Rica and Guatemala, but became a pest only in the Caribbean countries like St. Martin, Guadeloupe, Antigua, Bahamas, British Virgin Islands, Cuba, Dominican Republic, Haiti, and Puerto Rico and later from the Pacific regions of Guam and the Republic of Palau.

The papaya mealy bug invaded Florida in 1998 on hibiscus and within four years, by January 2002, it had been collected 80 times on 18 different plant species in 30 cities. The current status would include the whole of both Americas

The bug *P.marginatus* attacks hibiscus, pulses, tapioca, bhindi, pepper, tomato, tulsi, orchids, cotton, and many ornamentals and shrubs. Mulberry is also one of the hosts. Papaya, hibiscus, lablab and tapioca plants have perished due to the attack.

The damage is caused due to the sucking of sap from the veins of leaves of the affected trees, along with other parts like the fruits and flowers. The leaves become crinkled, lose their vigour, get yellowish, wither and then fall off. The adult female is yellow and is covered with a white waxy coating. Adult females are approximately 2.2 mm long and 1.4 mm wide. A series of short waxy caudal filaments less than 1/4 the length of the body exist around the margin. This differentiation is not easily understood by the ordinary farmer, but the presence of the cottony mass can give confirmation of the attack. The female bug lays up to 600 eggs within an ovisac and they start developing into the small babies or crawlers within 10 days. They move out actively from the mother ovisac and spread to the other parts of the same plant or other plants. Within a month, different stages are completed and the next cycle continues. So in one month, a single mealy bug can multiply to 600 numbers. The next month, even if half of them were males and incapable of egg laying, it would be 300 x 600, or 1, 80,000.This is not happening due to the active feeding by the natural enemies, which include the predators and parasitoids. When the temperature is favourable, in the absence of natural enemies, they increase tremendously. This is what happens now.

Biological control is a key component in a management strategy for the papaya mealy bug, and steps are on for classical biological control programs including importation of natural enemies and identification, mass multiplication and release of native natural enemies. They include the commercially available mealy bug destroyer (*Cryptolaemus montrouzieri*), coccinellids, lacewings, and hover flies, all which are generalist predators that have a potential impact on mealy bug populations. Larvae of the lycaenid butterfly, *Spalgius epius* is almost indistinguishable in the colony and is actively preying on the nymphs. There are some native parasitoids that were isolated, but not yet identified.

This year, the parasitoids *Acerophagus papayae* (Noyes and Schauff), *Anagyrus loecki* (Noyes and Menezes) and *Pseudleptomastix mexicana* (Noyes and Schauff) were imported from Puerto Rico. All three species were screened in quarantine facilities and environmental assessments were completed *Acerophagus papayae*has been identified as the most successful and is being distributed to the farmers in large numbers. The pest is now under control due the parasitoid action.

The more resourceful go for chemical control using insecticides, but even here, the doses required are very high. This is because of the waxy coating of the

mealy bug. Hence, it adds to the pollution of the environment and requires repeated application. Addition of powdered neem cake in the soil, and application of insecticides like Malathion (2ml/l) or imidacloprid (3ml in 10L), in mixture with liquid soap, (A shampoo sachet (8ml) suffices for two litre spray fluid) helps to dissolve the wax, and ensures better pest control. Since the eggs in the ovisac continue to emerge for another 10 days, it is necessary to give a repeat spray after 10 days. Addition of liquid soap is a must for better efficiency of spraying, since otherwise, the waxy coating prevents the insecticide from penetration on the pest.

Many wild plants are its hosts, and hence spraying cannot be fully effective. What are currently protected are only crops and ornamentals. Whole papaya trees have dried up due to the attack. But more damage to farmers would be due to the potential threat it has shown to tapioca, lab-lab, coconut, and black pepper.

Among pulses, the papaya mealy bug has become most serious on the dolichos bean and has caused heavy losses on the spreading type varieties, and has become very difficult to control also. Farmers must be advised to take preventive and prophylactic sprays at the initial attack itself. Mechanical control of all mealy bug stages is very important. It is also necessary to spray pesticides like Dichlorvos or Malathion, with less residual activity. But most importantly, the spray should be done against the crawlers when they are very small, and the spray should also contain liquid soap to digest the mealy coating without which the spray is ineffective. A dosage of 5ml of teepol per litre or any other liquid soap, or even ordinary shampoos that come in small sachets are sufficient. A second spray around ten days after the first is also needed to kill the next generation that springs up after the first spray.

Pod bugs: *Riptortus pedestris, Clavigralla horrens, Clavigralla gibbosa, Anoplocnemis phasiana*, Coreidae

Adults of Riptortus are dark brown elongated bugs with the hind femur conspicuously large and having a row of spines. The adult has two dark bands on the ventral side of the abdomen. The nymphs and adults suck the juice of unripe seeds from the green pods. In case of severe infestation the tender parts get shrivelled and later dry up. The bugs are seen clustered around on the pods.

Bionomics

Riptortus pedestris: The female bug lays an average of 115 eggs singly on pods at their base. The egg period is 3-4 days. The nymphs are dark brown and resemble the brown ant. The nymphal stage undergoes 5 instars completed in 16 days.

Clavigralla horrens is brown, flat, narrow - bodied bugs with conspicuous

lateral spines on the prothorax and enlarged hind femur with row of spines. They are 1.2 long, 0.4 mm wide-

C. gibbosa is smaller in size. *A.Phasiana* has a leafy hind leg.

These bugs lay eggs in groups of 3-15 on pods or leaves. The fecundity is 60-400 eggs per female. The incubation period is 4 days. There are 5 nymphal instars. The nymphal duration varies from 7-31 days. The adult bug lives up to 150 days.

Management

Spray the infested crop with 0.05 % endosulfan 35 EC or dust with 4% dust or Methyl parathion 2% dust.

Lablab bugs / stink bug: *Coptosoma cribraria*: Coremelaenidae

Symptoms of damage

Both nymphs and adults cluster on the tender shoots and pods and suck cell sap. Heavily infested shoots and pods wither and dry. Moderately infested plants remain weak and stunted in growth.

Bionomics

The adults are sub globular greenish bugs and have the characteristic stink bug odour when handled. They lay creamy white sculptured eggs in double rows in batches of 35-50 on the tender shoots and pods. Incubation period is about 7 days. Total life cycle as nymph and adult is about two months.

Management

When the attack is severe, dust the infested crop with any dust formulation that has low residual toxicity.

Green shield bug: *Nezara viridula* :Pentatomidae

The adults and nymphs are sap feeders. Adults are uniformly green small flat shield bugs1.3cm long.

Thrips: *Ayyaria chaetophora*: Thripidae

These are summer pests. The leaves are mottled with characteristic silvering due to the attack of insect especially under dry spell .The adults are tiny yellow thrips with fringed wings. Young ones are without wings. For managing the pest in severe form, spray endosulfan 0.05% or Imidachloprid 2.5-3ml/10 litre. Provide sprinkler irrigation or irrigate the whole plants just as a rain.

Whitefly: *Bemisia tabaci*: Aleyrodidae

Symptoms of damage

The damage is caused by both nymphs and adults, which are found in large numbers. The nymphs are closely attached to the leaves and hence are not easily seen. They also do not move. They suck plant sap and lower its vitality.

Severe infestation results in premature defoliation, development of sooty mould or honey dew and shedding of flowers and pods. Adults do feed on the plants, but the sap removal is very negligible.

Bionomics

Adult is a minute insect with yellow body covered with white waxy bloom. Nymph is greenish yellow, oval in outline along with puparia on the under surface of leaves.

Management

If the attack is severe spray Hostathion 40 EC 2.5-4ml/L, Imidachloprid 20 EC 0.5-one ml/L, Thiacloprid 0.5-one ml/L, Fipronyl5%2 ml/L or Endosulfan35EC 3ml/l. Provide sprinkler irrigation or irrigate the whole plants just as a rain.

Green leafhopper: *Empoasca kerri, Empoasca binotata*: Cicadellidae

Plants with heavy populations of nymphs show signs of hopper burn. The nymphs and adults feed on tender leaves and other parts by sucking the plant sap. In cases of severe attack, leaves become brittle and dry. The plants lose their vigour resulting in poor growth.

Elongate, active wedge shaped green insects are found on the under surface of leaves. The female inserts its eggs inside the veins of leaves. The incubation period lasts 4-8 days. There are five nymphal instars occupying 7-10 days.

This is a minor pest in cowpea normally managed by good irrigation and by the natural enemies. If the attack is severe spray Acephate 75 WP one gl/L, Imidachloprid20 EC 0.5-one ml/L, Thiacloprid 0.5-one ml/L, Provide sprinkler irrigation or irrigate the whole plants just as a rain.

Stem fly: *Melanagromyza phaseoli* :Agromyzidae

The adult is a small black fly. The pest damages cowpea When it is a seedling. The eggs are laid in cavities on the leaf or stem. Maggots mine the leaves or bore into the leaf petiole or tender stem. Larvae grow for 6-12 days, pupate inside the larval gallery and adult comes out in one to three weeks. The boring of shoots by the maggots cause withering drooping and death of seedlings affecting the crop stand. Increase the seed rate in endemic areas and apply insecticide granules at sowing. Seed treatment will also reduce the attack well.

Spotted pod borer: *Maruca testulalis* :Pyralidae

The adult is a slender moth 2.5 cm long. It is greyish brown having a large transverse hyaline band on the forewing and a few smaller spots. Hind wing is hyaline with brown outer margin.

The female lays eggs on or near the flower buds. The larvae bore the bud, flower or pod. The seeds inside the pods are eaten, while the entrance hole is plugged with excreta. The stems may also be bored. Flowers and pods can also be webbed. The full grown larva is 2cm long, brownish green and with black warts. Pupation is within the debris or near the surface of the ground. The moths remain among the lower leaves.

Spiny pod borer: *Etiella zinckenella*: Pyralidae

This is a small moth 1.7 mm long and with greyish forewings having the costal margin white and bearing a white transverse mark towards the base. Eggs are laid at the junction of the calyx and pod or on the pod surface. Up to 170 eggs are laid that hatch in 5-6 days. The larvae bore within the green pods and feed on the seeds. Larval period 10-13 days. They drop down to the ground when fully grown and form a cocoon an inch below the ground or under dry leaves. The pupae may take up to 20 days to become adults.

Plume Moth :*Exelastis atomosa* :Pterophoridae

The adults are small moths with fringed plume like wings. They lay minute eggs singly on young pods, flower buds or on tender leaves. The larvae emerge in about 3 days and bore into the pods. They feed on the seeds or bite into the unopened flower buds eating the developing anthers. The larvae do not enter he pods fully. They become full grown in about 25 days, are 1.3 cm long, greenish brown, and with fringed hairs and spines. Pupation is on the pod surface for about a week. The pupa resembles the larvae and has spine and hairs.

Pod caterpillar/borer: *Helicoverpa armigera*: Noctuidae

This is a medium sized moth yellowish brown in colour with awing span of about 3 cm and a body length up to 1.4 cm. There is a dark speck and a dark area on the outer margin of each forewing. The forewings also have greyish wavy lines and black spots of different sizes on the upper side. A black kidney shaped mark and around spot is seen on the underside of the forewing. The hind wings are lighter, whitish, with a broad blackish band along the outer margin.

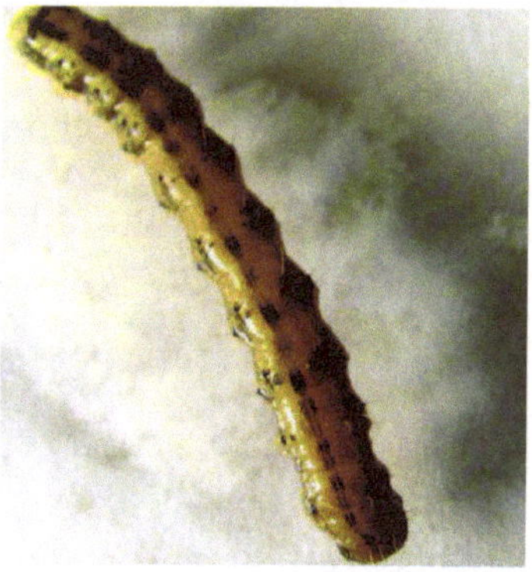

Damage

Young larva feeds on tender foliage and infests fruits from 4thinstar onwards. Make circular holes and thrust only the head and a part of its body into pod and eat the inner contents. The rest of the body is on the outside of the pod. A single caterpillar can destroy many pods.

The egg period is for 2-4 days, eggs being yellowish white, spherical and laid singly on tender plant parts. The larva live for 18-25 days, with the newly emerged ones being yellowish white and full grown being green. There are usually 5 instars and the larvae are cannibalistic in nature. They are fully fed by 2-4 weeks and are about 3-4 cm long. They are greenish brown with deep yellow stripes, but with many morphotypes based on the host plant and even the surroundings. White and dark grey longitudinal lines and sparse short hairs are seen in many morphs. Pupa takes 6-21 days and they pupates in earthen cocoon in soil. Total life period takes 4-6 weeks with 5-8 generations/year.

Management

- Collect and destroy the infected pods and grown up larvae.
- Grow simultaneously 40 days old American tall marigold and 25 days old tomato seedling at 1:10 rows to attract *Helicoverpa* adults for egg laying.
- Setup pheromone trap with Helilure at 15/ha and change the lure once in 15 days.
- Six releases of *T. chilonis*@ 50,000/ha per week coinciding with flowering time and based on ETL.
- Release *Chrysoperla carnea* at weekly interval at 50,000 eggs or grubs / ha from 30 days after planting.
- Spray HaNPV or commercial formulations of NPV.
- Spray endosulfan 35 EC 2 ml/lit or carbaryl 50 WP 4g/lit or *B. thuringiensis*2 g/lit or quinalphos 2.5 ml/lit.
- Do not spray insecticides after maturity of fruits.
- To encourage activity of parasitoid *Eucelatoria bryani*, *Campoletes*, *Chelonus* etc., use less residual insecticides and prolong the interval between sprays.

Pod borer: *Adisura atkinsoni*: Noctuidae

This is a stout built pale yellowish brown moth. The adult lays the eggs on the flowers which are developing and on tender pods. The larvae bore the pods and feed on he seeds. The fully fed larva is brownish green, and about 2.5 cm in length. Pupation can be on the flower spikes or also in the soil. Egg to adult is completed in about a month.

Pod borer: ***Lampides boeticus*** **: Lycaenidae**

The adult is a beautiful blue butterfly of medium size, one cm long and with about 2.5 cm wing expanse. They lay their eggs on flower buds and pods. The larvae are flat and pale green or violet in colour. They burrow into the buds or tender pods, feeding on them. pupation is on the plant itself. The larvae may be attended by ants and has the habit of remaining partly exposed at the opening of the bore hole on the pod.

Leaf eating caterpillar: ***Spodoptera litura*****: Noctuidae**

Damage: The adult is a stout greyish brown moth 3.8x1.8 with wavy white markings and criss-cross lines on the forewings and white hind wings having a brown patch along its margin. Young larva scrape leaves on ventral surface, and growing larvae defoliate crops and also feed on young fruits.

Eggs are laid in masses covered with white hairs on the surface of affected leaves. Eggs hatch in 4-5 days. Larva eats for 14-20 days. They are pale brown with greenish to violet tinge and with stout whitish or creamy bands running the length of the body. Yellow and purplish spots found in sub marginal areas. Faecal pellets are seen on the leaves and on the ground which is the indication of the pest incidence Pupa is seen in the soil for 7-15 days.

Spodoptera litura

Management

- Plough the soil to expose and kill the pupae.
- Grow castor along borders or irrigation channels as indicator or trap crop.
- Flood the field to drive out hibernating larvae.
- Set up light trap at 1/ha or pheromone trap at 15/ha with spheroclin SL lure.
- Destroy egg masses and grown up larvae.
- Spray SlNPV at 250 LE along with teepol 1ml/lit in evening hrs.
- Spray Carbaryl 50 WP 4 g/L or Fipronyl 2ml/L or Dichlorvos 76 WSC 1 ml/l or endosulfan 35 EC 3ml/l or NSKE 5% preferably in the evening before the larvae set off for feeding.

Leaf folder: *Anticarsia irrotata*: Noctuidae

Adult is yellowish brown moth with oblique black lines on the wings. Larva is green coloured. The Larva folds the leaves together.

Epilachna beetles :*Afidenta misera*: Coccinellidae

This is one species of coccinellids that exactly resembles the common epilachna species but is taxonomically different. The biology, damage and management are almost the same for all the species. The classification with the number of spots, is not accepted by taxonomists. According to them, the species on pulses is *Afidenta* and not Henosepilachna or Epilachna.

Lab-lab leaf miner: *Cyphosticha coerula*: Gracillariidae

Tiny larvae bore into the epidermis of the leaf and forms blisters through mining.

Termites: *Odontotermes obesus*: Termitidae

Termites damage by eating the seeds that are sown and continues to feed on the seedlings and bigger plants. The leaves of damaged plants droop down, which later wither and dry. Such plants are easily uprooted. This will prevail if the field is dry.

Where the pest is of regular occurrence, the soil should be mixed with 10% Lindane dust @ 35 kg/ha or 4% Endosulfan dust or Methyl parathion 2% dust.

Sphingid caterpillar: *Acherontia styx*,: Sphingidae

Lepidoptera

Adult is large, wings grey with waxy markings abdomen crimson coloured with black stripes. Larva is a stout green caterpillar with yellowish oblique stripes with curved anal horn. The giant sized larva feeds on leaves and cause severe defoliation of even whole plants unless removed by hand and killed.

Bruchid: *Callosobruchus chinensis*: Bruchidae

Damage is mainly as a grain borer in storage. But infestation starts from field. Seeds are eaten by the grubs from inside and turned hollow. This is the most serious storage pest.

Leaf-eating caterpillar: *Spilosoma obliqua* (Bihar hairy caterpillar, Arctiidae); *Euproctis fraterna* : Lymantriidae

Large numbers of hairy caterpillars are defoliators of cowpea and other pulses. They destroy leaves initially by scrapping and later by eating the whole lamina.

IPM for Cowpea

It is a good practice to go for seed treatment if there is incidence of stem fly and termites. In the nursery stage, for sap feeders, a spray of any insecticide like Imidachloprid20 EC2.5-3ml/10l, Thiacloprid 0.5-one ml/l, Fipronyl5% 2ml/l or Endosulfan 35EC 3ml/l may be necessary if the pest attack is noticed. Observe for the presence of pests and give the sprays in the initial stages of the attack itself, to prevent pest build up.

During the growth stage, sucking pests and pod borers are important. When pod borer incidence is noticed, sprays with insecticides like carbaryl 50 WP @ 4gm /l or malathion 50 EC 2ml/l or endosulfan35 EC 3 ml/l or quinalphos 2 ml/l or thiodicarb75 WP @1.5gm/l can be used in the beginning of the season. During plucking stage, use only Malathion, or dichlorvos or other low residual insecticides. These would also manage other shoot borers and leaf feeders. When dusting is needed, especially against pod bugs, 4% Endosulfan or Methyl parathion 2% or carbaryl 5 or 10% dust or other dusts may be used, just enough to ward off the pests.

Chapter - 75

CUCURBITS

Fruit flies: *Bactrocera cucurbitae*:Tephritidae

This pest is more serious in bitter gourd, snake gourd and to a certain extent on ridge gourd. Other cucurbits are attacked to a lesser level. Damage by this pest causes losing the fruit quality, rotting and fruit drop. The adults lay many eggs inside the fruit sing their big ovipositor that hatch out into worm like apodous maggots. Feeding by the maggots on the pulp of fruits result in rotting, yellowing and fruit drop. The dropped fruits allow pupation by the maggots in the soil from where they emerge out as adults and start egg laying on fresh frits.

Adult of *B. cucurbitae* have hyaline wings with a thick band on the costa and prominent well developed apical and anal patches and spots. *B. ciliatus* is smaller than *B. cucurbitae* with ferruginous brown body and a prominent dark brown oval spot on either side of 3rd tergite. *B. cucurbitae* is the prominent species.

Management

- It is a must to collect infested fruits and dried leaves and **burn** or bury in deep pits.
- In endemic areas, the sowing dates are to be adjusted to get the fly population low in hot dry conditions and avoid its peak during rainy season.
- Expose the pupae by ploughing and turning over soil after harvest.
- Use kairomone traps like coconut shell trap using crushed mashed ripe banana pulp as the lure. The pulp can be kept in a coconut shell or other container and a pinch of furadan granules added to it. Sprinkle water over the contents every day and remove the dead flies. Change the fruit pulp when it loses the attractiveness. Up to 30 traps can be hung per hectare of crop. The traps can be hung on the periphery of the trellis for crops like bitter gourd or snake gourd, and can be kept on raised areas protected from rain, in crops like pumpkin and cucumber.

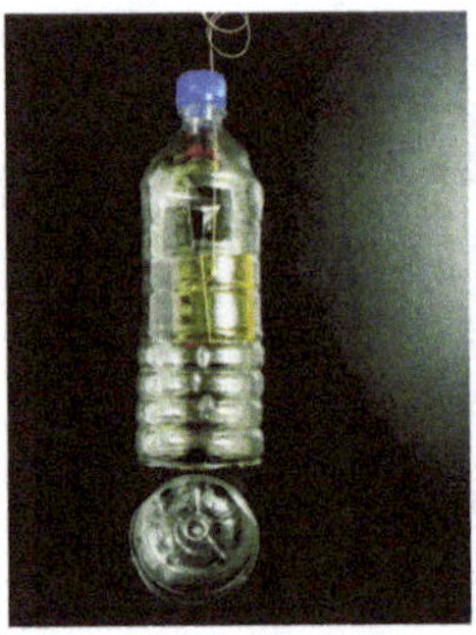

- Coconut shell trap MAT trap Use male annihilation technique using methyl eugenol impregnated plywood traps, with malathion added as the toxicant.
- Give bait sprays on the crop and on the undersurface of neighbouring plants with a bait spray containing sugar@10g/l plus malathion 4ml/l.
- Use attractants like citronella oil, eucalyptus oil, vinegar (acetic acid), dextrose and lactic acid to trap flies.

Caution

In cucurbits, DDT, Lindane, Copper oxychloride, Bordeaux mixture and Sulphur dust should not be used as these are highly phytotoxic. If needed, spray 50 WP carbaryl 4gm/l or malathion 2ml/l

Pumpkin beetles/Leaf beetles:*Aulacophora foveicollis, A. cincta, A.intermedia*: Galerucidae

Adults are medium sized, 6-7 mm long. *A. foveicollis* is red, *A. cincta* is grey with black having glistening yellow border and *A. intermedia* is blue in color. The adult feeds on leaf and flowers, and grubs feed on the roots, stem and fruits touching the soil. Grub, when freshly hatched is dirty white and full grown ones are creamy yellow, around 22 mm long.

Management

i. Early planting of pumpkin to avoid peak periods of occurrence.

ii. Frequent raking of soil beneath the crop to expose and kill the eggs and grubs.

iii. Hand collection and destruction of infested leaves and fruits.

iv. Spray malathion 50 EC 1 ml/lit, dimethoate 30 EC 1.5 ml/l or dichlorvos one ml/l .

Aulocophora foveicollis

Aulacophora stevensi

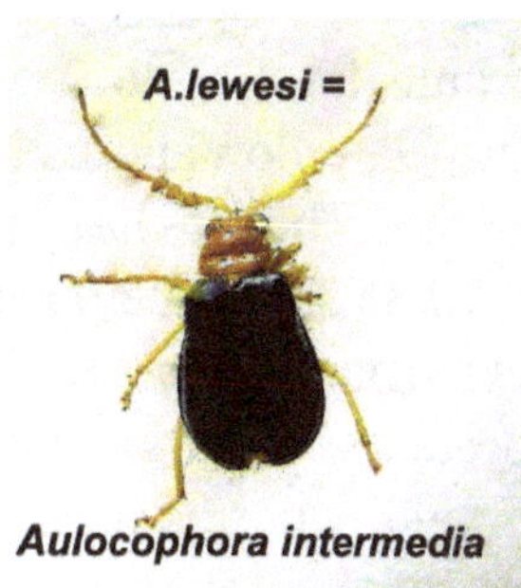

Aulocophora intermedia

Snake gourd semilooper/Leaf eating caterpillar: *Anadevidia (=Plusia) peponis*: Noctuidae

The adult is a stout dark brown moth with dark patches on the forewing. The wing span is about3.7 cm. Eggs are laid on the tender leaves and the green semiloopers have black warts on the body. The caterpillar cuts the edges of leaf lamina, folds it over the leaf and feeds from within leaf roll. Pupation is in the leaf folds itself. They could fully denude young snake gourd plants, flowers and fruits, but are usually parasitized heavily .For management, spray malathion 50 EC 1 ml/l, dimethoate 30 EC 1.5 ml/l or dichlorvos one ml/l.Care should be taken to harvest the fruits and then spray. Systematically collecting and destroying the larvae itself can manage the pest to a great extent.

Epilachna beetle: *Henosepilachna septima*:Coccinellidae

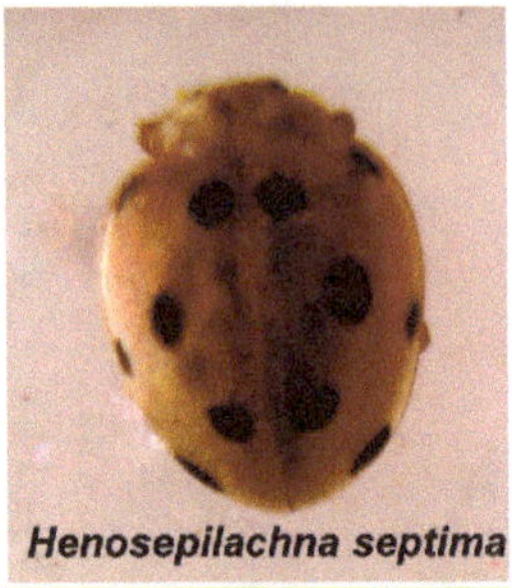

Both adults and grubs scrape the lower epidermis of leaves in characteristic manner leaving behind riddled areas. The leaves give a grazed appearance. In severe infestation all leaves may be eaten off leaving only the veins intact (skeletonization), with the leaves or whole plants withering.

Adults lay 120-460 eggs/female. They are cigar shaped, yellow and are laid in clusters on lower leaf surface. The eggs hatch in 2-4 days. The grub period is 10-35 days. They are yellowish bearing six rows of longitudinal spines with terminal branching. Pupa occupies 5-6 days and is seen on the stem or leaves. Total life period occupies 20-35 days with 7 -10 generations / year.

Management

- Systematically collect and **destroy** leaves with egg masses, grubs and pupae and also adult beetles
- Spray carbaryl 50% WP 4 g/l or endosulfan 35 EC 2 ml / l or malathion 50 EC 2 ml /l when populations explode. Diflubenzuron or Buprofezin also works well on the beetles

Leaf miner :*Liriomyza trifolii* :Agromyzidae

The larval feeding causes serpentine mines on the leaf surface. Drying and dropping of leaves is seen due to severe infestation.

For pest management, application of NSKE 5% or 50g/lit as required is a good practice.

Note: *HCH (BHC), DDT should not be used in cucurbits. Copper and sulphur dusts may be phytotoxic to gourds. For management, remove and destroy severely affected leaves. spray the leaves with quinalphos or other contact insecticides early in the morning to kill the larvae that move out to mine fresh areas or to pupate in fresh areas.*

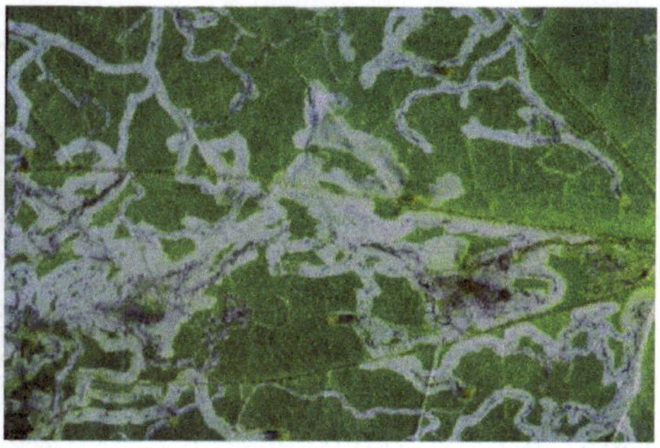

Aphids :*Aphis gossypii* :Aphididae

The aphids are light greenish to greenish brown depending upon the stage of growth. They congregate on the lower surface of fresh leaves in the initial stages and later spread to other parts. Attack by the winged and wingless forms cause desapping, crinkling and loss of growth. Nip the attacked leaves regularly and destroy them. When the attack becomes severe, apply Malathion one ml/l or dichlorvos one ml/l

Leaf hopper:*Hishimonus phycitis* :Cicadellidae

Nymphs and adults of this leafhopper desap from the undersurface of leaves and cause crinkling and yellowing. The nymphs and adults are yellowish green. Severe attack causes browning and drying of the leaves that starts from the margins of the leaves and can cause upward curling. If needed, apply neem oil emulsion with soap solution @ 5ml neem oil plus 5 ml liquid soap per litre of spray solution. Severe attack can be prevented by applying insecticides like malathion one ml, dimethoate 2ml or carbaryl 4gm/l.

Plume moth: ***Sphenarches caffer*** **: Pterophoridae**

The adults of this pest are tiny plume moths with feathery thin wings. They lay the eggs on the buds and leaves. The larvae grow up to 7 mm, is yellowish green and with spines all over the body. They pupate on the leaves and the pupa is also spiny. These are important pests in bottle gourd.

Pumpkin caterpillar: ***Diaphania(=Margaronia) indica*** **:Pyraustidae**

The adults of this pest are black bordered on both the wings with hyaline central areas and a tuft of orange hairs at the abdominal tip. They are small, measuring 2.5 mm in wing span and 1.3 mm in length. Eggs are laid on the underside of leaves that hatch in 3-4 days. They bind the leaves together and feed within. The caterpillars are bright green with a pair of thin white longitudinal lines running on the middle of the dorsal side. There is a cocoon for the pupa that is seen among the leaves. Adults emerge in about a week. The attack causes leaf, flower and fruit damage to melons, cucumber, and gourds. Destroy eggs and larvae periodically. Insecticides that can be used for management include malathion one ml, dimethoate 2ml or Dichlorvos one ml or carbaryl 4gm/l.

Red pumpkin bug ***Coridius (=Aspongopus) janus, C.brunneus, C.observus*****: Dinidoridae**

Coridius (=Aspongopus) janus, is a large shield bug 1.8x1.1cm with the dorsal side red and black and the ventral side, head and legs black. This is seen to suck the sap from plant parts of pumpkin. This pest also is not serious.

Spiralling whitefly: ***Aleurodicus dispersus*** **:Aleyrodidae**

This ubiquitous pest is also a pest of many cucurbits, feeding from the underside of leaves, and from other plant parts. The pest is so named because of the specific nature of the pest which lays the eggs in a spiral mass. Immature and adult stages of *A. dispersus* cause direct feeding damage by sucking plant sap, which can cause premature leaf fall. Indirect damage is due to the heavy production of honeydew and white, waxy material produced by the insect. Sooty mould develops on honeydew and decreases photosynthetic activity. Plants are also disfigured and the fruits become unmarketable. In places where it occurs, *A. dispersus* is generally considered as a serious pest, causing crop losses. Virus transmission is apparently not known. When the crop is well cared for and other chemical treatments are given to the plants, they do not attain pest status. This is an introduced pest that was very serious during 2000-2005 , and now is managed by the native and introduced parasitoids and predators. If the attack is serious, a spray on the underside of the leaves with malathion 0.05%, Dichlorvos 0.05% or other low residual insecticides has to be done. Add liquid soap@5 ml per l to the spray solution. Repeat the spray one week later to manage the new generation of crawlers.

Blister beetle: ***Mylabris pustulata*** **: Meloidae**

This is one of the common insects that show hyper metamorphosis. There are different shapes for the different instars. The adult feeds voraciously on buds and flowers. A single beetle can destroy as many as 20-30 flowers/day. The eggs are laid by female beetle in clusters of 60-80 eggs at 2-3 cm depth in soil. Eggs are light yellowish in colour and cylindrical in shape. Incubation period is about three weeks. Young grubs are white in colour. The first instar is a triungulin. Then the first grub phase comes in. In the sixth or seventh instar, the coarctate phase is seen and /or the second grub phase. Then it pupates inside the soil tunnel and adult emerges.

Stem gall fly : *Neolasioptera falcata*: Cecidomyidae

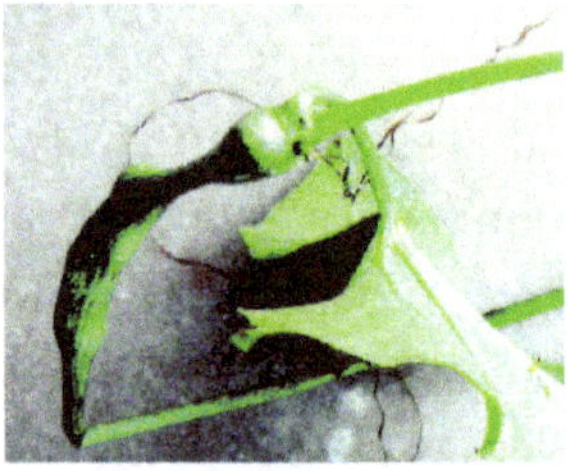

This is a gall fly. The eggs are laid on the vine tips. They hatch out and cause the production of galls at the terminal ends of shoots in bitter gourd and Ivy gourd/coccinia. Growth of the vine stops due to the attack. Recently it has become a serious pest in bitter gourd and coccinia(Ivy gourd).Cut the vine tips with the galls and burn or crush them. If the attack persists, spray a contact insecticide to kill the adults. This author has also observed fruit flies emerging from the galls, but further confirmation is awaited.

Snake gourd bug: *Leptoglossus australis* :Coreidae

The adults of this bug are black and about 2cm long. The eggs are laid in a row, each row having 16-17 eggs. There are five nymphal instars that last for about two months. The nymphs and adults feed on the sap of tender vines and causes their death, or drying of leaves or fruits or malformed fruits.

Thrips: *Thrips tabaci, Frankliniella sulphurea* : Thripidae

Adults and nymphs of these thrips feed on tender leaves which later curl towards the inside.

Non Insect Pests

Spider mites:*Tetranychus cinnabarinus (=T. urticae= T. telarius = T. cucurbitacearum); T. neocaledonicus*: Tetranychidae

These are carmine red coloured small arthropods that move actively on the under surface of leaves. They spin webs and live in colonies beneath the webbing. Continuous feeding by the mites result in development of yellowish spots on the feeding sites which later turn brownish. Mites damage the leaf by scratching the leaf surface by their mouth parts and sucking the exuding sap. The leaves lose their turgor and the photosynthetic efficiency is lost and later the leaves dry up fully.

Management

Application of wettable sulphur 80wp 2g/l or Dicofol or Dimethoate 30 EC1.5ml/l or Phosalone 2ml/l on the under surface of the leaves. Provide copious irrigation on the under surface of the leaves.

IPM in cucurbits

Cucurbits, as in the case of other vegetables, have to be protected from insects, with minimal use of insecticides. Mechanical removal of egg masses, larvae, pupae and adults is the rule. Insecticides, when inevitable, should start with neem based insecticides. In extreme cases, need based use of dichlorvos, malathion and quinalphos may be made. It should also be remembered that in cucurbits, DDT, Lindane, Copper oxychloride, Bordeaux mixture and Sulphur dust should not be used as these are highly phytotoxic.

Chapter - 76

MUSHROOM

Sciarid(mushroom flies): ***Bradysia paupera, Lycoriella auripila*****: Sciaridae**

The larvae feed on the mycelial attachments and restrict the growth of the spawn. The maggots also feed on the developing mushrooms or **pinheads** and thus cause loss in yield. They also bore the developing mushrooms and render them hollow.

Phorid (dung flies): ***Megaselia halterta, M. agarica, Megaselia*** **spp. Phoridae**

The female lays eggs in the base growing medium or directly in the gills of opened mushrooms. The larvae feed through the bottom or directly into the cap from the gills. The infested mushrooms become yellowish brown and then stop growing.

Staphylinid beetle :*Staphylinus* spp.: Staphylinidae The grubs of this beetle feed on the soft gills of mushroom

Mite(mushroom mite):*Tyrophagus putrescentiae*: Acaridae

These are translucent white with long hairs on the body. They feed on mycelium and damage the sporophores. The caps shrink and develop brown rusted spots. They can also kill young buttons.

Mycetophagous mites: *Pygmephorus* spp. Pyemotidae

The mites feed on the mycelium.

Nematodes: ***Aphelenchoides composticola, Ditylenchus myceliophagus*****: Tylenchidae**

These nematodes feed on the mycelium, which becomes sparse and patchy and the compost colour turns brown. Flushes become poor and delayed with decline in the yield.

IPM in Mushroom

1. **Sciarid flies, Phorid flies and cecidomyid flies**
   a. Do not keep the spent compost and casing material at moist and shady places as they serve as ideal substratum for fly breeding during non cropping period.
   b. Dispose the material in the manure pit and cover it with at least 10cm thick layer of manure to check fly breeding in surrounding areas of the mushroom house.
   c. Cook-out of the spent compost and casing material at 71 °C with steam for 2 hours to kill immature stages of the fly in spent material.
   d. Put screen for doors and ventilating openings of the mushroom house with nylon wire nets of 14-16 mesh/cm to check the entry of flies into the mushroom house.
   e. Incorporate chlorpyriphos 50 ppm (50 mg/l) in the compost and casing material to reduce larval population in the beds and fly population in the mushroom house.

Chapter - 77

TOMATO

Fruit borer: *Helicoverpa armigera*: Noctuidae

This is a medium sized moth yellowish brown in colour with a wing span of about 3 cm and a body length up to 1.4 cm. There is a dark speck and a dark area on the outer margin of each forewing. The forewings also have greyish wavy lines and black spots of different sizes on the upper side. A black kidney shaped mark and a round spot is seen on the underside of the forewing. The hind wings are lighter, whitish, with a broad blackish band along the outer margin.

Young larva feeds on tender foliage and starts infesting the fruits from later instars. They make circular holes and thrust only the head and a part of its body into the fruit and eat the inner

Helicoverpa armigera

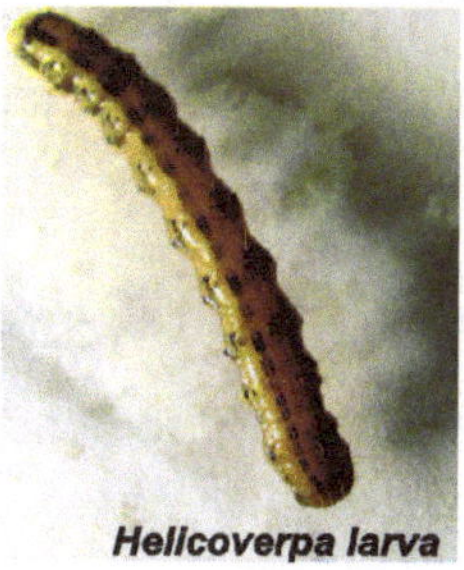

Helicoverpa larva

contents. The rest of the body is on the outside. A single caterpillar can destroy many fruits

Spherical yellowish white eggs are laid by the female singly on tender plant parts. Eggs hatch in 2-4 days. Larvae eat for 18-25 days with newly emerged coloured yellowish white and full grown with dark broken grey lines on the side of the body. They have 5 instars. The larvae are cannibalistic in nature and can feed on one another when food is scarce. They are fully fed by 2-4 weeks and are about 3-4 cm long. They are greenish brown with deep yellow stripes, but with many morphotypes based on the host plant and even the surroundings. White and dark grey longitudinal lines and sparse short hairs are seen in many morphs. Pupa: 6-21 days pupation is in earthen cocoon in the soil that takes about 6-21 days. The total life cycle occupies 4-6 weeks with 5-8 generations / year.

Management

- Collect and destroy the infected fruits and grown up larvae.
- Grow simultaneously 40 days old American tall marigold at 1:10 rows to attract *Helicoverpa* adults for egg laying.
- Setup pheromone trap with Helilure at 15/ha and change the lure once in 15 days.
- Conduct six releases of *T. chilonis* @ 50,000/ha per week.
- Release *Chrysoperla carnea* at weekly interval at 50,000 eggs or grubs / ha from 30 days after planting.
- Spray HaNPV or commercial formulations of NPV.
- Spray malathion 50 EC one ml/l or quinalphos 2 ml/l.
- Do not spray insecticides after maturity of fruits.

To encourage activity of parasitoid *Eucelatoria bryani*, *Campoletes*, *Chelonus* etc., use less residual insecticides and prolong the interval between sprays.

Serpentine leaf miner: *Liriomyza trifolii* : Agromyzidae

This is an introduced pest that got into India in the last decade due to deficient quarantine. Their maggots mine into leaves and causes serpentine mines, and later drooping and drying of leaves.

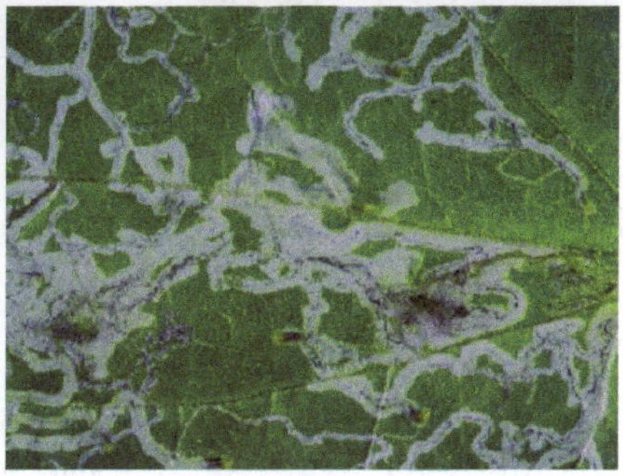

The incidence of leaf miner is a major problem in all tomato growing areas both in the open and in green houses. The attack is severe in March-June. The flies lay eggs in the leaf tissues. The maggot just after hatching makes narrow, linear mines in the leaf by feeding the mesophyll tissues. In severe cases leaves show extensive mining which appear as zigzag silvery lines on the upper surface of leaves. The infested leaves turn yellowish, then browning and later dry up.

The female thrust eggs into the epidermal layer of leaves. The egg period is 2-4days.Larva develop inside for 7-10 days. They are minute orange yellowish apodous maggots. Pupation is within mines and takes about 5-7 days. The adult is pale yellow in colour.

For management, it is important to monitor the pest occurrence regularly. On sighting the first of the leaf mines, they have to be systematically collected and destroyed by thrashing the leaves. Apply NSKE 3% or other neem formulations, by dusk or at least very early in the morning when the maggots cone out of the mined leaves for getting into fresh areas. When severe, remove the affected leaves , destroy them and spray dichlorvos, malathion or other insecticides that have less than three days of residual action.

Leaf eating caterpillar: ***Spodoptera litura*****: Noctuidae**

The adult is a stout greyish brown moth 3.8x1.8 with wavy white markings and criss-cross lines on the forewings and white hind wings having a brown

Thick line and shadowy spread around this line along its margin. Young larva scrape leaves on ventral surface, and growing larvae defoliate crops and also feeds on young fruits.

Eggs are laid in masses covered with white hairs on the surface of affected leaves. Eggs hatch in 4-5 days. Larva eats for 14-20 days. They are pale brown with greenish to violet tinge and with stout whitish or creamy bands running the length of the body. Yellow and purplish spots found in sub marginal areas. Pupa is seen in the soil for 7-15 days.

Management

- Plough the soil to expose and kill the pupae.
- Grow castor along borders or irrigation channels as indicator or trap crop.
- Flood the field to drive out hibernating larvae.
- Set up light trap at 1/ha or pheromone trap at 15/ha with pheroclin SL lure.
- Destroy egg masses and grown up larvae.
- Spray SlNPV at 250 LE along with teepol 1ml/lit in evening hrs.

Spray Quinalphos 25 EC 2ml/lit, or Malathion 2ml/l or Dichlorvos 76 WSC 1 ml/l or Endosulfan35EC 3ml/l or NSKE 5%, spray insecticide during evening hrs.

The Whitefly: *Bemisia tabaci*: Aleyrodidae

Adult is a minute insect with yellowish white body and a white waxy bloom for the wings. Nymph is greenish yellow, oval in outline.

Nymph

Adult

The damage is caused by both nymphs and adults, which are found in large numbers. The nymphs are closely attached to the leaves and hence are not easily seen. They also do not move. They suck plant sap and lower its vitality. They also secrete honey dew and heavy population will aid development of sooty mould. Severe infestation results in premature defoliation, development of sooty mould and shedding of flowers and pods. Nymph is greenish yellow, oval in outline and are seen along with puparia on the under surface of leaves.

Management

If the attack is severe spray Imidachloprid20 EC 0.5-one ml/L, Thiacloprid 0.5-one ml/L, Fipronyl5%2 ml/L, Endosulfan35EC 3ml/l. Provide sprinkler irrigation or irrigate the whole plants just as a rain.

Thrips: *Thrips tabaci*; *Calliothrips indicus* : Thripidae

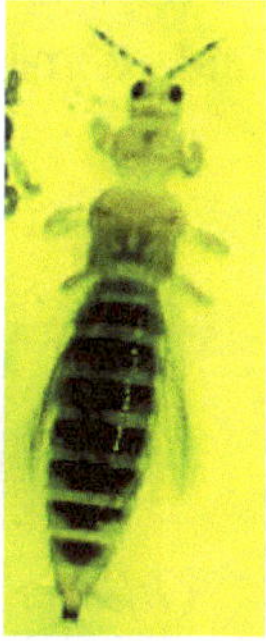

Damage

This is the vector of tomato spotted wilt virus. The adults and nymphs lacerate leaf tissues and leaves become spotted and pale with silvery streaks. They also feed on flowers resulting in pre-mature dropping of flowers and also bud necrosis.

Biology

Thrips have an egg period of 3-5 days and nymphal duration of 5-10 days. There is a stage that is referred as pupa that lasts for about 6 days. The adult is a minute, fragile slender insect with wings that are fringed with hairs. Total life cycle for one generation can occupy 14-32 days depending on the temperature, with higher temperature favouring lower life period for attaining adulthood and higher number of eggs and more generations in a crop period. Females live longer than male and there can be several overlapping generations /year.

Management

- Collect and destroy the damaged leaves and twigs
- Use yellow sticky traps at 15/ha to attract and kill insects.
- Release first instar larvae of *Chrysoperla carnea*@ 10,000/ha and encourage coccinellid predators.
- Spray Quinalphos 25 EC 2ml/lit, or Malathion 2ml/l or Dichlorvos 76 WSC 1 ml/l or Endosulfan35EC 3ml/l or NSKE 5%, to control vector.
- Spray insecticide during evening hrs.

Fruit sucking /piercing moths: *Othreis materna*, *O. fullonica*, *O. ancilla*: Noctuidae

These moths also attack citrus, mango, grapes and apple. The adult moths pierce the fruits for sucking the juice and make characteristic pin-hole damage in fruits. Bacterial and fungal infections take place at the site of attack. The whole fruit turns yellow due to the sap sucking, and it drops down. In severe cases, many fruits can be lost, but these are usually only minor pests. Adults of *O. fullonica* has a wing span of 9-10 cm and are grey with white patches and a demarcating line from anal tip to base making a triangular division of the

Othreis materna

O. fullonica

O. ancilla

wing. Hind wing has a black patch on the outer margin and a crescent moon black patch facing the outer edge in the middle of the yellow wing.*O. materna* have white and dark patches on the greenish brown forewings and a black spot on the hind wings. *O. ancilla* has white bands in the middle of the grey forewing with whitish outer margins. Hind wings are yellow with an outer black margin and a central black spot.

Eggs are laid on wild plants and weeds in and around citrus and tomato orchards. The egg period lasts for about two weeks. They are round and translucent. Eggs are usually laid on weed plants like Rangoon creeper (*Quisqualis indica)*. Larvae are semi-loopers and have a stout appearance having a dorsal lump on the last segment of the body. They have distinct yellow / red eye spots on head/ lateral sides. They have 5 instars completed in about 4 weeks. Pupa is in a pupal case by webbing together pieces of leaves and soil particles. Pupa is thick and dark reddish brown.

Management

i. Destroy the weed hosts like Rangoon creeper.

ii. Set up food lures with rotten tomatoes (or) pieces of citrus fruits baited with a contact poison like malathion 0.03%.

iii. Cover the fruit with perforated poly bags.

iv. Bait with fermented molasses / jaggery + malathion 1 ml/l.

Epilachna beetle/Hadda beetle: *Henosepilachna viginti octopunctata, H. dodecastigma* :Coccinellidae

Adults of these species were distinguished with the number of spots on the elytra. *E. vigintioctopunctata*: 14 spots on each elytra, deep red. Adult of *E. dodecastigma*: Copper-coloured, 6 spots. Now, taxonomists say that those spots are not distinguishing character.The one on cucurbits is *H.septima*, on brinjal it is *H.viginti octopunctyata* and on pulses, it is *Afidenta misera*

Both adult and grubs scrape the lower epidermis of leaves in characteristic manner leaving behind riddled areas. The leaves give a grazed appearance. In severe infestation all leaves may be eaten off leaving only the veins intact (Skeletonization), with the leaves or whole plants withering.

Henosepilachna septima

Adults lay 120-460 eggs/female. They are cigar shaped and yellow eggs are laid in clusters on lower leaf surface. The eggs hatch in 2-4 days. The grub period is 10-35 days. They are yellowish bearing six rows of longitudinal spines with terminal branching. Pupa occupies 5-6 days and is seen on the stem or leaves. Total life period occupies 20-35 days with 7 -10 generations / year.

Management

- Systematically collect and destroy leaves with egg masses, grubs and pupae and also adult beetles.
- Spray carbaryl 50% WP 4 g/l or endosulfan 35 EC 2 ml / l or malathion 50 EC 2 ml /l when populations explode.
- Diflubenzuron or Buprofezin also works well on the beetles.

Leafhopper: *Amrasca biguttula biguttula (=A.devastans)*: Cicadellidae

Adult is a greenish, slender, boat shaped leafhopper. It lays eggs singly within leaf veins. Incubation period is 4-6 days. The nymph is yellowish green found on the under surface of leaves. The nymphal period is 7 days. The life cycle is completed in 15-46 days. Both nymphs and adults suck the plant sap from the under surface of leaves. Tender leaves become yellow and start crinkling. They also start curling downwards and reddening sets in. In the case of severe infestation leaves get a bronze or brick red colour which is typical "hopper burn" symptom. The margin of the leaves get broken and crumble into pieces when crushed. The leaves dry up and are shed and the growth of the crop is retarded. The injury is also due to the toxaemia produced by the toxic saliva that the hoppers inject into the plants. The plants stop growing and no fruits are produced

If the infestation is severe, spray quinalphos @ 2ml/l or Malathion one ml/l or NSKE 5% @ 25 kg/ha or endosulfan 3 ml/l .

Mealy bugs: *Ferrisia virgata* :Pseudococcidae

This is also called the two tsailed mealy bug. The white cottony mealy bugs are seen on plants congregating on the shoots and leaves.. The nymphs and adults suck plant sap and stunt growth of plants.

Bugs: *Nezara viridula* : Pentatomidae

Sometimes green flat bugs gregariously feed on tomato leaves.

Mite: *Tetrancychus cinnabarinus* : Tetranychidae

Small red mites seen generally on lower side of leaves. They cause webbing and feed from below.

IPM in Vegetables

Reducing insecticide application to the lowest level is the motto of IPM in vegetables. Monitoring for the initial build up and timely mechanical removal of the first population is a must in vegetables. Always remove the first affected plants, for diseases like mosaic, and destroy them by burying or burning. Remove insect eggs, insect affected plant parts and adults and immature stages of insects, and destroy them every day. Do not throw them in the field. They will either come back, or pupate and attack later. During nursery stage, use wood ash, cow dung solution, and plant extracts to deter all pests. Use only low residual insecticides like Dichlorvos 1ml/l, Malathion 2ml/l, Endosulfan 2ml/l or Quinalphos 2ml/l, in spots, at the beginning of attack itself. Use pheromone traps against fruit and shoot borer in brinjal, bhindi, diamond back moth in cabbage and cauliflower etc.. Set up sticky yellow traps against aphids and whiteflies in cowpea, brinjal, bhindi, cabbage and cauliflower. Whereever feasible, grow trap crops like mustard in cabbage fields for monitoring, or acting as a barrier or also for attract and kill strategy.Always grow resistant varieties and treat the seeds with seed treating dusts (STD) such as Imidachloprid, before planting. Release parasitoids wherever available, like *Chelonus blackburni* @ 4000/acre in bhendi, *Trichogramma chilonis* @ 20000/ acre/week against Helicoverpa and *Cotesia plutellae* against Diamond Back Moth.

Nematodes: ***Aphelenchoides composticola, Ditylenchus myceliophagus*****: Tylenchidae**

These nematodes feed on the mycelium, which becomes sparse and patchy and the compost colour turns brown. Flushes become poor and delayed with decline in the yield.

IPM in Mushroom

1. Sciarid flies, Phorid flies and cecidomyid flies:

   a. Do not keep the spent compost and casing material at moist and shady places as they serve as ideal substratum for fly breeding during non cropping period.

   b. Dispose the material in the manure pit and cover it with at least 10cm thick layer of manure to check fly breeding in surrounding areas of the mushroom house.

   c. Cook-out of the spent compost and casing material at 71 °C with steam for 2 hours to kill immature stages of the fly in spent material.

   d. Put screen for doors and ventilating openings of the mushroom house with nylon wire nets of 14-16 mesh/cm to check the entry of flies into the mushroom house.

   e. Incorporate chlorpyriphos 50 ppm (50 mg/l) in the compost and casing material to reduce larval population in the beds and fly population in the mushroom house.

Chapter - 78

IPM IN VEGETABLES

Reducing insecticide application to the lowest level is the motto of IPM in vegetables. Monitoring for the initial build up and timely mechanical removal of the first population is a must in vegetables. Always remove the first affected plants, by diseases like mosaic, and destroy them by burying or burning. Remove insect eggs, insect affected plant parts and adults and immature stages of insects, and destroy them every day. Do not throw them in the field. They will either come back, or pupate and attack later. During nursery stage, use wood ash, cow dung solution, and plant extracts to deter all pests. Use only low residual insecticides like Dichlorvos 1ml/l, Malathion 2ml/l, Endosulfan 2ml/l or Quinalphos 2ml/l, in spots, at the beginning of attack itself. Use pheromone traps against fruit and shoot borer in brinjal, bhindi, diamond back moth in cabbage and cauliflower *etc*. Set up sticky yellow traps against aphids and whiteflies in cowpea, brinjal, bhindi, cabbage and cauliflower. Always grow resistant varieties and treat the seeds with seed treating dusts (STD) such as Imidachloprid, before planting. Release parasitoids wherever available, like *Chelonus blackburni* @ 4000/acre in bhendi, *Trichogramma chilonis* @ 20000/ acre/week against Helicoverpa and *Cotesia plutellae* against DBM.

FSC
www.fsc.org
MIX
Papier | Fördert gute Waldnutzung
FSC® C083411

Zeitfracht Medien GmbH
Ferdinand-Jühlke-Straße 7
99095 Erfurt, Deutschland
produktsicherheit@kolibri360.de